NUCLEAR POWER — THE UNVIABLE OPTION

WITHDRAWN

To Nancy Ann Guinn
for the love that helped me through this book
and for her sound editing

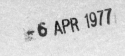

NUCLEAR POWER
THE UNVIABLE OPTION

A Critical Look At Our Energy Alternatives

John J. Berger

Foreword by Senator Mike Gravel
Introduction by Dr. Linus Pauling

Ramparts Press
Palo Alto, California 94303

Library of Congress Cataloging in Publication Data

Berger, John J
  Nuclear power — the unviable option.

  Bibliography: p.
  Includes index.
  Contents: Gravel, Senator Mike Foreword — Pauling,
  Dr. Linus Introduction —
  1. Atomic Power. 2. Energy conservation. 3. Renewable
natural resources.   I. Title.
TK9153.B47   621.48   76-2181
  ISBN 0-87867-062-9
  ISBN 0-87867-063-7 pbk.

Published by Ramparts Press, Inc., Palo Alto, California 94303

Library of Congress Catalog Card Number 76-2181

Printed in the United States of America.

Contents

Part I Nuclear Power

Part II Clean Energy Alternatives

Part III The Rising Tide

ILLUSTRATIONS

FOREWORD

In the last two years, nuclear power has become a political issue of national prominence. Voters across the country are demanding that they have a say in America's decision to "go nuclear."

And this is happening none too soon.

At first, only a handful of concerned citizens and scientists dared to challenge the "Atomic Establishment," a multi-billion dollar alliance of government and industry, inflated with tax-payer dollars and protected by the military secrecy of the Atomic Energy Commission. Today, this "Establishment" faces a nationwide co-alition of nuclear power opponents, the strongest grass-roots move-ment since the Vietnam War.

These nuclear opponents have recognized that atomic power is not a narrow issue of energy supply, nor is it even confined to the dynamics of energy/economy/environment. They see that nu-clear energy poses a broad social and military threat—the threat of totalitarian society and the ever-increasing chance of nuclear war — even beyond the extraordinary dangers of reactor accident and radioactive waste management.

Yet, even with the nuclear debate at a crescendo, the subject remains complex and difficult for the newcomer. The need for energy often seems more apparent than the nuclear threat — and when confronted with the question of nuclear power, most people are still likely to say, "Well, why not?"

This book tells why not.

It describes accidents which have already occurred at U.S. reactors — Browns Ferry, Fermi, SL-1 — and it describes the kind of catastrophic accident we have so far been spared. It outlines the continuing spillage and leakage of radioactive wastes and the government's inability to implement a final waste management program. It explains why nuclear reactors are not covered by full liability insurance — and it shows how the nuclear power industry came into being only because the government agreed to suspend normal liability requirements for reactors.

The book lays out the possibility of irreversable radioactive contamination, and it explains how nuclear pollution could threaten the genetic integrity of life on earth. It shows why our civil liberties would probably be curtailed to keep plutonium out of the hands of terrorists. And it tells why the proliferation of peaceful nuclear reactors will lead to the proliferation of nuclear weapons.

The answer of nuclear proponents to the host of problems besetting nuclear power is, "We need the energy." And yet, the kind of crash nuclear program advocated by successive Administrations would mean *less* energy, dollar for dollar (and fewer jobs, as well) than we could expect from increasing end-use energy efficiency, in the short term; and, in the long term, from solar energy.

Nuclear energy, with its huge initial reactor costs, aggravates the energy capital shortage; domestic resources of uranium are inadequate to fuel a nuclear economy without the breeder reactor; and it now appears that, in addition to its great new dangers, the breeder will hardly "breed" new fuel at all.

Many utility companies have recognized that nuclear power is a dead end. New reactor orders in this country have come to a stand-still.

But our government's energy agencies still push, to the tune of a billion dollars a year, for nuclear power. Why?

The unavoidable conclusion, it seems to me, is that civilian nuclear power grew with the atomic bomb. The Atomic Energy Commission was anxious to assure a broad technological and man-power base for its weapons program. In addition, feelings of guilt over having used atomic weapons impelled many in government to seek a peaceful use for fission power.

Today, the atomic industrial base is huge; ongoing nuclear research contracts are as sweet as defense contracts; and there is an institutional momentum seeking the profits which goverment (not the marketplace) continues to assure.

Only a broad-based citizens' movement can stop the government-supported momentum of nuclear power. Nuclear opponents must make clear to others a threat which is more dangerous than that of the Vietnam War — and more difficult to percieve.

This book explains why that task must be achieved.

Senator Mike Gravel (D-Alaska)

INTRODUCTION

The United States and the world as a whole are in poor shape. Part of the reason for this is the waste of a large share — 10 percent — of the wealth of the world and the products of man's labor on war and militarism. This burden has now reached $300 billion per year, 60 percent of it wasted by the United States and the Soviet Union.

War and militarism involve the misuse of energy, but energy is also misused in many other aspects of our civilization. It is difficult to foresee the future, but I can predict that the immediate future is bleak. I am afraid that within twenty-five years or fifty years there will occur the greatest catastrophe in the history of the world. It might well result from a world war, which could destroy civilization and might be the end of the human race: or civilization might be destroyed and the human race brought to an end because of the collapse of the natural systems upon which it depends. The end of civilization might result from changes in the weather induced by governments to improve the yield of crops, or it might end by the rapid destruction of the ozone layer in the stratosphere, or by the accumulation of poisonous wastes that would make the air unbreatheable and water undrinkable.

Another possibility is that the construction of more and more electric power plants depending upon nuclear fission would lead to a catastrophe. The plan now is to construct immense nuclear power plants by the hundreds. I believe that no more nuclear fission power plants should be constructed. The reasons for my belief are essentially those presented in this book, *Nuclear Power: The Unviable Option*, by John J. Berger. I hope that every responsible citizen will read this book and think about the problem.

There are now about eighty nuclear fission power plants in the United States, and they already constitute a serious problem, associated especially with their safety and the danger presented by the tremendous amounts of highly radioactive fission products that they produce. No one knows how to handle these fission products in such a way that future generations of human beings will be safe, and no one knows how to construct nuclear power plants free from the risk of a catastrophic melt-down that might kill hundreds of thousands of people.

We may ask how we got into this situation. The decisions have been made for the most part by the leaders of nations and governments in general, and by the people who run the banks and the large corporations. Many of these people, of course, make their decisions in accordance with the profit motive. Moreover, these decision makers, including the leaders of nations and governments, concentrate on the immediate problems. It is unusual for a country to develop a five-year plan, and unheard of to have a hundred-year plan. Only when a crisis arises, when a catastrophe occurs, do governments take action.

The crisis has arisen now — the crisis of the nuclear fission power plants. If the present policy of building them in larger and larger numbers is not stopped, the world may well be changed in such a way as to make it impossible for future generations of human beings to lead good lives.

I believe that decisions among alternative courses of action should always be made in such ways as to minimize the predicted amount of human suffering. I believe that the goal for the United States of America should not be to remain the strongest military power in the world, or to achieve an increase of gross national product by 7 or 8 percent each year. Instead, I believe that the goal that we should strive to reach is that of constructing a country and a world in which every person has the possibility of leading a good life.

A good life is one in which there is a minimum of suffering and a maximum of satisfaction. To minimize suffering we must provide every person not only with adequate food, clothing, and shelter, but also with education to the extent that he or she can benefit from it, with the opportunity to develop to the fullest extent. Freedom of choice in personal actions is essential; also the preservation of different cultures, which enrich the world, and the preservation of the world's natural wonders, the redwood forests, the wildernesses, the mountains, the lakes, which should not be sacrificed to the advancing technology.

The problem presented by the continued increase in the world's population is a serious one. Education and the provision of incentives for limiting the number of progeny are needed. It is inevitable that the population will reach a limit at some time during the next century. The population of the world, now four billion, is expected to be eight billion within thirty-five years, and this might constitute the limit. The limit would be reached when the maximum exploitation of the earth's capability for producing food has been achieved by sacrificing forest lands and other natural resources that should be conserved. At that point, even the discoveries of our chemists and other scientists yielding increased supplies of edible and nutritive materials would have become insufficient to prevent the death by starvation of millions of people. The alternative that we should strive to achieve is to limit the population to the value that would permit every person, including those of future generations, to lead a good life and to experience the minimum amount of suffering.

Most of the people in the world do not lead good lives now. One-third of the people in the United States are poor, and two-thirds or more of the people in the whole world.

The energy available to persons determines in part whether or not they can lead good lives, but only in part. The people of the United States use twice as much energy per person as the people of Sweden, but I think that no one who knows the two countries could say that on the average better lives are led by the people in the United States than by those in Sweden. Much of the energy that we use is wasted. In this book, in the chapter on conservation, the ways to reduce the energy use per person without a reduction in the quality of life are discussed.

The generation of greater and greater amounts of energy is damaging the earth. The principle of minimizing the amount of human suffering needs to be applied not only to the people who are now living but also to future generations. Future generations should not be deprived of their share of the world's natural resources.

Some of the world's natural resources are inexhaustible, and these are the ones that we should be exploiting, so as not to diminish the wealth of the world and to cause unnecessary suffering for future generations of human beings. An example is sunlight. We may convert some of the energy of sunlight into electric power, which, after doing useful work, is finally converted to heat and warms the earth, as it would have done if it had not been converted to electric power. The sun will continue to shine for future generations, whether we use it for electric power or not. The use of solar energy — like that of energy obtained from the winds or tides, also coming largely from the sun — does not rob future generations.

Coal and oil, however, are present on earth in definite amounts, and when they are burned the supply is decreased. These are valuable substances, not only as fuels but also as raw materials for the chemist to use in making drugs, textiles, and many other compounds of carbon. But we are depleting their stocks at such a rate that they may be exhausted in a few generations.

The use of nuclear fission for the generation of electric power is not the solution, because there is only a limited supply of fissionable elements, and our extensive development of fission power plants would soon exhaust the nuclear fuel, leaving little or none for future generations.

I advocate that no more nuclear fission power plants be constructed, and that instead we concentrate our efforts on decreasing the waste of energy and other resources on militarism, decreasing the waste of energy on trivialities, decreasing the waste of energy through inefficient operations, and developing supplies of energy from inexhaustible sources.

The immediate step to be taken is that of stopping the construction of nuclear fission power plants.

I believe that, by considering the problem and by making the proper decisions, those that will lead to the decrease in the amount of human suffering, we can achieve a world in which every person can lead a good life.

— Linus Pauling

Part I

NUCLEAR POWER

1

THE SANTA BONITA SCENARIO
A NUCLEAR PHANTASM

Santa Bonita was a picturesque central California town of 19,840 when chosen in 1965 by the San Philipe Municipal Utility District as the site for a nuclear electric generating station. The town had a lovely, eighteenth century Spanish mission and was situated in a fertile agricultural valley near the Delta Mendota Canal and the huge San Luis Dam. San Philipe engineers had been quick to see that a large nuclear plant could draw cooling water from either source, and that its big generators would not be more than fifty miles in a beeline from the heavily populated San Francisco Bay Area to the north and the scenic Monterey Peninsula to the south. Santa Bonita was also within forty miles of three fast-growing California cities — San Jose, Merced, and Modesto.

Those areas, the San Philipe planners had reasoned, would be needing vastly expanded power supplies if electricity consumption grew at its 7-percent-a-year historical rate. So they decided to build three jumbo-sized pressurized water reactors at Santa Bonita, making the new plant the largest in the state. At 3.9 billion watts, it would have more than enough power for a city of two and a half million that depended exclusively on the plant for electricity. Although the growth in electricity consumption had slackened as the plant neared completion, the utility district was financially so committed to the project that abandoning the plant was out of the question. Besides, company planners had reasoned that by the time the plant was done, other electric utility companies would offer to buy the surplus power.

After a ten-year planning and construction period, over the objections of environmentally concerned groups, and after many mechanical difficulties, the plant finally began full power operation.

As usual with nuclear plants, it was gingerly started up at low power and was carefully tested as the power level was increased. Much to the company engineers' and contractors' satisfaction, the plant went on line with nary a hitch. Moreover, ultrasonic inspection of reactor piping attested to a high standard of manufacturing and assembly quality control: far fewer defective components were found at Santa Bonita during the first year of operation than were normally discovered during routine inspections at reactors of this type. Although the nuclear industry had begun to acknowledge that ultrasonics sometimes failed to detect known leaking cracks in stainless steel pipes,[1] this very testing process was relied on for most pipe inspections at Santa Bonita. Thus, although all inspectors performed their tasks conscientiously at Santa Bonita, following all inspections procedures, and although manufacturers' radiographic inspections were all honestly performed and reported, the project engineers' reliance on ultrasonics proved to be the chink in Santa Bonita's safety armor. This was later confirmed in the extensive official post-mortem on Santa Bonita.

*   *   *

After five and a half years of successful full-power operation, Santa Bonita encountered its first serious operating difficulty — one capable of prematurely aging the plant's circulatory system. The plant had just reached full power again after a refueling when a water chemistry imbalance was detected. Water chemistry must be carefully controlled by demineralizers in a reactor to minimize pipe scaling and corrosion that could reduce a pipe's resistance to stress. An excess of free dissolved oxygen in the water, for instance, could cause the formation of oxides on the insides of pipes. For reasons which the company's engineers were not immediately able to pinpoint, excessive chlorides and silicates were turning up in water samples, and plant engineers were concerned about their effects upon critical regions of the plant's steam supply system, such as the thin-tubed internals of the plant's steam generators. Harvey Kuen, the plant manager, had just instructed his staff that the reactor would have to "go down" later that week unless the water imbalance could be eliminated. It was not long after this decision that a tragedy occurred.*

Randall McFarr, fifty-six, the senior control room operator, was on duty at the Santa Bonita plant, assisted by Gene Yardley, another qualified operator. Several equipment operators and technicians were also at their jobs throughout the plant. But the onus for preventing the catastrophe fell on McFarr, a man whose nerves and health had been severely strained by more than a decade at high-pressure jobs in major U.S. power plants.

The previous night, McFarr had slept poorly as usual. When he caught sight of a large split-second dip in reactor Unit 1's flow rate gauge (measuring reactor coolant flow), McFarr wasn't sure if his eyes were deceiving him. "If that's what I think it was," he said to himself, "we've got another screwy gauge on board today. A miracle I noticed it at all," he mused, surveying the hundred-odd other gauges in the room. "Must be getting psychic, McFarr, after all these years." Then a queasy feeling niggled at the pit of his stomach for the briefest instant, and he rubbed a palm heavily against his forehead: What if the meter had been telling the truth?

If the coolant flow had really suffered a major arrest, why hadn't the warning light flashed? Why hadn't the alarm sounded? Had Kagan reconnected the meter properly after the test recalibration? The reactor should have automatically "TRIPPED" if the flow change was genuine and not an instrument error or a result of that sleepless night.

"TRIP" meant immediate shutdown, a process Santa Bonita operators tried to avoid except when a problem threatened public safety or imperiled the integrity of their two-billion-dollar plant. Shutdowns were expensive, of course, and could not be initiated on flimsy provocations. McFarr understandably felt hesitant to put the reactor into a shutdown mode as yet. If anything serious went wrong, the reactor was designed to shut itself off automatically and to tell him why, via flashing lights and gauge readings.

Walking over to the suspicious dial, he stared at it and tapped it with his knuckle, but it was reading properly once again. "You're not givin' me a hard time today, are you?" he said to the gauge.

---

*In creating this scenario, we will mention some technical systems that may be unfamiliar to the reader. Their functioning will be clear from their context, and the most important of these reactor systems will be explained more fully in Chapter 2.

Yardley looked up with a start. "Begad," said McFarr, parodying himself as he often did to relieve tension, "I think we've just had a coolant flow change." McFarr flipped on the public address system without waiting for Yardley's reply.

"LeBaron, Kagan," McFarr said glumly into the microphone.

"What can I do you for?" a disembodied voice replied almost immediately. McFarr's control panel showed Kagan was at Station 12, near the plant's steam turbine outside the containment building for reactor Unit 1. As McFarr spoke, he could imagine the rangy technician leaning his back against the plant's huge outdoor crane and gazing at the massive slope of the giant Westinghouse turbine that had taken twenty-one flatcars to haul to the site. Although McFarr was a qualified reactor operator and had a video scanner with which he could view the vitals of the plant, he was also an operator of the old school who had been in the industry since the opening of Dresden I and Yankee Rowe in 1960 and 1961. Where other operators preferred their instruments to fallible human observations, McFarr had a firm belief in the importance of personal inspection permitting manual examination of equipment and visual inspection of systems. He just liked to "eyeball things," McFarr would say, "at least until someone installs a 3-D color video scanner up here in my control room."

"Got a job for you," McFarr said now to the twenty-five-year-old equipment operator. "I know you're on lunch, but something's on the kibosh up here. We've got an oscillating 130 over 120 reading on the flow rate meter with a slight downward list. Check the main coolant pumps and the steam generators. Our meter readings here need to be confirmed."

McFarr personally had hoped that forty-eight-year-old Stan LeBaron would have been free. Compared to LeBaron, McFarr felt that Kagan was still wet behind the ears. McFarr never quite trusted the young man because Kagan had leapfrogged from junior college straight into Westinghouse's reactor operator school without cutting his eye teeth on a conventional fossil-fueled steam plant, the way LeBaron had.

After putting down the phone, equipment operator Kagan went into a locker room and donned his white radiation protection coveralls and then headed for the reactor containment entry airlock. Inside the containment building, the steel steps of the first walkway and then of the fuel loading bridge high above the reactor vessel rang eerily underfoot.

Beneath the bridge, Kagan could survey the guts of the reactor building — a complex array of pumps, generators, pipes, valves, and the reactor vessel itself — pressurized to 2,250 pounds per square inch. Behind him stood freezer compartments, nearly sixty feet tall. In the event that a major reactor pipe leak were ever to occur in the sealed containment building, the condensors were intended to cool the escaping reactor steam by passing it over ice. In reducing the outward steam pressure against the containment building dome, the condensors were supposed to help prevent containment bursting. But Santa Bonita was one of the first reactors ever to actually be built with ice condensors, and they had never been tried in an emergency. Above Kagan's head in the containment dome loomed the silhouette of the building's 125-ton crane, used for reactor refueling.

He climbed down a ladder from the catwalk amidst the subdued thunder of giant machinery. Nothing looked amiss to the casual eye — it rarely did — but Kagan had a vague premonition that something wasn't right. Even after two years of duty at Santa Bonita, Kagan still didn't really feel comfortable right next to the reactor itself. The noise in the containment coupled with its pressure differential made his head feel uncomfortable. As he negotiated the steep stairs from one elevation to another, he felt slightly lightheaded, without quite knowing why. For an instant, he even had a vague impulse to call in sick, but he had gone to the bother of going this far, and he didn't want to quit.

Pausing for a moment in his descent, he chanced to glance up at the containment dome, now hearly a hundred feet above his head and, for some inexplicable reason, he thought of the accident at the SL-1 experimental reactor in Idaho. An instructor at reactor training school had mentioned it. During the accident, a technician had been blown to the ceiling of the containment dome where he had hung impaled on a control rod for six days before reactor experts and workers could get his body down. Both of the technician's coworkers had been killed. "Hey, Kagan, don't scare yourself," he thought, "you know that only happened once."

Kagan descended another set of stairs. He could feel them tremble under his feet, but he could no longer hear them over the noise of machinery. Finally, he stood like an ant on the concrete containment floor amidst the maze of enormous reactor piping.

To Kagan's right towered one of the plant's steam generators, a steel hull seventy feet high containing thousands of tubes about half an inch thick, each individually welded to a tube sheet. These tubes contained a lot of surface area susceptible to stress-induced cracking or to leakage as a result of water chemistry imbalances, not to mention various kinds of thermal and mechanical shocks that could result in the rupture of thin-walled tubing. That was why the tubes at Santa Bonita and elsewhere were forever being taken completely out of action by plugging them to prevent damaged tubes from leaking.

Kagan now stood near one of the plant's primary coolant pumps, which was handling about 88,000 gallons of hot presssurized water per minute, circulating water through the reactor to cool the core and transfer its heat to the generator to make steam. In the center of the reactor building and in front of Kagan was the reactor vessel itself — more than four hundred tons of it. Its studs and nuts alone weighed twenty tons.

Kagan was moving away from the pressurizer and toward the first pump when he thought he heard a faint hiss, much like that of air entering the vacuum seal of a coffee can. During the next three seconds, the thought "Weak weld" careened through Kagan's mind. By reflex action and curiosity, he turned towards the sound. As he did so, a three-inch bypass line suddenly split with a sharp backfire. Instantly, the pipe began flailing violently around as though the finely tempered steel were nothing more than paper, whipped by the enormous forces of the superheated pressurized water geysering from it in an ever-expanding cloud of radioactive steam.

Kagan leaped toward the stairway as if his whole body had been jolted by an electric shock, but before he could even get his feet on the bottom step, the searing vapor enveloped him in a dense cloud, scalding his entire body. Screaming, he raced, stumbling up the stairs, tearing off the water-soaked clothing in which he was literally being boiled alive.

An instant after the pipe break, a shock wave from the fracture point traveled faster than an eye could see back along the line, actually rippling the metal pipe with energy. When the shock hit the thin tubing inside the steam generator, six of its tubes simultaneously ruptured.

For the next fifteen seconds, everything at Santa Bonita began working just as the textbooks and the government-sponsored safety studies said they should. Automatic sprays on the inside of the containment building began condensing the billowing steam to prevent overpressurization that could rupture the containment. The plant's high pressure injection system phased in twenty seconds after the break, but although it slowed the reactor's depressurization, it rapidly became apparent that it was unable to stop the pressure loss.

Upstairs in the control room, it seemed to Yardley and McFarr that every alarm in the plant had gone off in their ears at once. "We've got a LOCA [*loss of coolant accident*]," McFarr shouted first, and immediately slammed his fist on the reactor's shutdown switch, just in case automatic shutdown was not already in progress. Yardley, meanwhile, leapt to the video screen and scanned the interior of the containment building, but all he could see on his monitor were clouds of steam. "Kagan's still in there!" Yardley yelled. Everything now seemed to be happening at once. In addition to the clamor of alarms, the control room's telephone lines all lit up. Simultaneously, both operators were frantically scanning the banks of control room gauges, trying to deduce exactly what had gone wrong.

In the midst of all this, an auxiliary control room operator burst into the control room in response to the general plant alarms, and he, too, began scrambling from gauge to gauge. Reading one of the scores of signals from the plant's pressure transducers, the auxiliary operator shouted, "The level's too low!" and at that moment, McFarr thought again of Kagan and began grabbing for his telephone, punching phone buttons desperately looking for the one call that might be from the containment entry-way. For security reasons, no single person could enter or leave the containment without the assistance of a control room operator. Kagan, if alive and functioning, would be frantically trying to gain his attention from the airlock in order to leave the reactor building.

On the second line McFarr picked up, his eardrum nearly split. It was Kagan screaming frantically with all his might, "Get me out!" McFarr could hear hysterical sobs in between inchoate phrases and suddenly, just as he was reaching for the containment door release switch, he had a vision of Kagan, coated with radio-activity, racing madly through the plant's corridors, rushing towards the control room.

"C'mon, Billy, what happened?" McFarr demanded, "are you hot [radioactive]?"

"I'm burned!" Kagan cried, "Burning all over!"

"The ambulance is coming, Billy," McFarr said, "we'll get you right away." He grabbed the emergency phone.

Yardley did not even wait for him to complete the call to the hospital. "Not enough backup coolant's reaching the core!" he said with a hoarse cry. "The control rods are in but the primary system's dry — the emergency coolant isn't reflooding. Look at the temperature!" McFarr turned his head and felt his throat constrict as he looked at the temperature arrays. Most of the gauges were already off the high end of the scale.

Within the twenty-five minutes after the pipe break, the reactor's pressure fell below 600 psi, and the depressurization triggered the one-way check valves of the acculator injection system. Driven by nitrogen gas, this emergency device suddenly began dumping thousands of gallons of borated water into the reactor cooling system.

In response to alarms, the plant supervisor and half a dozen terrified plant employees, against regulations, were gathering inside the double doors of the control room.

The two operators stared at their instruments and struggled to provide adequate coolant flow to the core, praying that any second the crucial instruments would return to normal. It was an uphill fight to operate the machinery while responding to a barrage of suggestions from the shift supervisor and others. Suddenly Yardley cried, "The temperature's climbing again. The emergency coolant's definitely blocked!"

"It's blowing out the break!" another voice cut in. "We've got to cool the thing some other way!"

"God damnit, we're doing what we can!" McFarr shot back. "Call the sheriff's office. Tell them we've got a bad LOCA!"

As soon as the call went through appropriate channels, a general evacuation order went out from the area's Office of Emergency Services to the population surrounding the plant. By now, several company nuclear experts had responded to the situation at

Santa Bonita and all the engineers were frantically trying to improvise a way to cool the phenomenally hot reactor and to prevent a breach of containment.

Despite the possible impending melt-through, a special ambulance team trained in handling contamination victims raced to Santa Bonita with its sirens blasting to pick up the young technician. He was already in shock and the burned skin of his arms and thighs was hanging in flaps and shreds from his body where it had come off as he tore away his clothes. The ambulance team spirited him off in a "contamination carrier" to a nearby hospital where he was admitted by an outside elevator directly to the autopsy room of the morgue for decontamination. A special team trained in nuclear medicine wearing full radiation protection gear strove to save him, but he only survived a day before succumbing to his burns.

Within the reactor building, although the pipe break had caused the reactor's shutdown program to activate automatically and the emergency cooling system also had phased in automatically —triggered by temperature, pressure, and water level sensors — the reactor was still in serious trouble. Core reflooding to prevent overheating and meltdown still was not proceeding properly. The steam generator tube ruptures had caused highly pressurized steam to leak into the path of the incoming emergency cooling water, preventing its entry into the reactor core. Consequently, despite the reactor shutdown, overheating continued from fission product decay while thousands of gallons of newly injected coolant roared out through the broken pipe instead of entering the reactor vessel. The water began filling the containment building sumps as the 350,000 gallon water refueling storage tank poured out its reserve of cooling water.

In less than an hour, the coolant had been dispersed into the reactor building where it was collected from the floor by drains leading to the containment sump. Inside the reactor vessel, fuel rods had begun melting in two minutes after the core emptied, and the rods' molten zirconium cladding reacted with water in the bottom of the vessel, releasing explosive hydrogen. It was now especially crucial for the containment spray to continue functioning to hold down steam pressure and pressure from exploding hydrogen, in the event the gas were to ignite. Because the refueling water storage tank was depleted, it was now up to the men in the control room to realign the emergency cooling water pump suction from the tank to the containment sump, to continue the spray action.

Nuclear Regulatory Commission analysts who attempted to reconstruct the accident later are not sure whether at this point operator McFarr activated the wrong valve sequence by pushing the wrong buttons or whether a valve jam interfered with the transfer of suction from the spent refueling water tank. For whatever reasons, the containment spray water failed to begin its recirculating cycle and containment pressure began soaring.

While operators in the control room worked frantically to improvise makeshift cooling arrangements for the reactor and struggled to keep the melting core from leaving the reactor vessel, accident processes inside the reactor vessel had entered an irreversible phase.

During the hour after the vessel had depressurized, the melting reactor core holding a hundred tons of fuel had been dripping onto the reactor's lower grid plate. When most of the fuel had melted, the plate suddenly buckled, dropping eighty tons of molten fuel into the pool of water contained in the reactor bottom head. Exposed to this gigantic thermal shock, the water expanded suddenly in a tremendous steam explosion that blew huge quantities of core radioactivity out of the reactor vessel and into the containment atmosphere. About that time, a spark from a sensor ignited ambient hydrogen with a powerful concussion that added hydrogen explosion pressures to the containment atmosphere in which steam pressures had already mounted in the absence of any containment spray. As the building pressure reached 100 psia, the pressure forced open a six-inch crack in the containment building ceiling through which radioactivity began leaking out of the building like air from a punctured balloon. Propelled by the containment overpressure, poisonous vapors laden with a witches' brew of radioactive elements surged into the atmosphere, where at the height of the crack in the dome, a strong wind of thirty-three miles an hour was blowing.

Within the reactor vessel, only a tenth of the searing hot molten fuel remained after the explosion. These ten tons of uranium, along with some zirconium and steel, lay directly against the reactor vessel, heating it like a super-energized blowtorch. Inexorably, the vessel began glowing red and buckling.

In the control room, most of the crowd had fled when sensors had registered the effects of the steam explosion. Now only the supervisor, Yardley, and McFarr were left as radiation alarms on their panels signaled the release of radioactivity from the containment. Both men were now wearing radiation protection suits. "Let's get out of here!" Yardley yelled at McFarr. The senior operator shook his head and stood grimly at the main control panel, feeling a sense of responsibility for the accident, and yet baffled and terrified by it. Although he did not know what to do now, he hoped nonetheless for a miracle and remained at his post as Yardley raced from the room.

Within another hour, the fiery mass of fuel had heated the bottom of the reactor to a white heat and liquid steel was dripping from the bottom of the vessel, releasing an ever larger mass of the molten core fragment directly onto the floor of the containment building. The hot puddle that formed there quickly cracked and broke the concrete, which crumpled and bubbled, releasing carbon dioxide. While this core residual was burrowing its way beneath the containment building, the invisible radioactive cloud that had already escaped from the containment building through the ceiling had risen to an altitude of roughly fifteen hundred feet as it spread in a plume downwind of the reactor. Drifting west with the wind, the cloud soared through the Pacheco Pass and across the Diablo Range.

On the other side of this natural barrier, the cloud was caught by a gentle, southerly breeze while a thermal inversion kept it from rising into the upper atmosphere. The radioactive mass, by then several miles wide, continued north, following the natural contours of the Santa Clara Valley.

As the cloud passed through the valley's lush agricultural regions, it coated everything with a radioactive film of lethal contaminants. People and animals in its path — from livestock to insects and aquatic life — all absorbed varying doses of the poison. For some humans, the effects were negligible, perhaps only to be noticed in the birth defect of a future descendant several generations later, or not at all, as chance might have it. For others who received higher doses, the results were more immediate and sometimes calamitous, depending on the amounts of radioactivity to which the victims were exposed.

As soon as civil defense workers had begun spreading the evacuation order, alarmed citizens ran for their cars to flee the silent menace. But as with most human undertakings, the evacuation was not faultless. Some individuals managed to escape notification of the evacuation until it was well under way. Others refused to heed reasonable warnings until they could assure themselves of their families' safety. But the bulk of the affected population began heading onto the freeway and most people in the plant's immediate vicinity went south, away from the radioactivity's predicted path. Communities more than half an hour's drive north of the plant were advised to head farther north to stay beyond the advance of the radioactivity. Just as the evacuation seemed to be progressing smoothly, however, the wind suddenly changed direction for about twenty minutes, and then reverted to its original course. This was enough to confuse people and to cause the civil defense office to begin issuing conflicting escape route directives. Many refugees now became unsure which path the poisons were likely to be taking, and many of the people south of the plant, who would have been safe had they followed the command's original instructions, now fled into the radioactivity's path, where they helped create impenetrable traffic jams.

Within thirty minutes after the first warning was issued, masses of motorists were clogging the four-lane blacktop, which was the only main thoroughfare leading away from Santa Bonita. Mobility was further impaired by a head-on collision that blocked two of the four available lanes. Many people in the sweltering bumper-to-bumper traffic thoughtlessly sat with their car windows open as toxic radioactive breezes imperceptibly swept in.

As it traveled away from Santa Bonita, the radioactive cloud settled on homes, roads, businesses, fields, freshly plowed topsoil, and growing crops, The fallout also landed on hospitals, schools, and government buildings, as well as on the Coyote and Anderson Reservoirs.

While the Santa Bonita region mobilized, panic spread north to the sprawling San Jose area, which also lay within the accident's official fifty-mile fatal dose range. From the moment the Santa Bonita containment burst, it took the radioactivity only about an hour and a half to engulf southern San Jose.

Air Force jets sent to measure airborne radioactivity levels with radiation monitors radioed frightening conclusions to ground control: a large percent of the Santa Bonita's core radioactivity had been spewed out in the accident. This datum, combined with information from ground radiation monitors, led authorities to fear that thousands of square miles of California surrounding the Santa Bonita site might have to remain a no man's land for the foreseeable future.

To prevent citizens from returning to the contaminated area to retrieve their possessions or, in a few cases, to loot, martial law was clamped on California, and all roads leading into the Santa Bonita region were sealed. This enabled the governor to use federal forces to patrol the state's new internal borders.

Tens of thousands of troops were summoned, and they attempted to cordon off central California from Santa Rosa to Sacramento on the north in a giant 170-mile long rectangle that stretched south to an arbitrary line connecting Monterey and Fresno. San Jose, San Francisco, Oakland, Stockton, Modesto, Salinas, and Merced all lay helplessly in its confines. The state suddenly had become trizonal, almost bisected by the disaster region. Intrastate commerce overnight fell into disarray.

Military decontamination and medical aid teams were dispatched to stations bordering the quarantined area. There, dressed in radiation protection garb, they monitored radiation levels on people fleeing the danger zone and conducted elementary decontamination procedures. Exposed people whom the military were able to intercept were given showers and their clothing was confiscated along with any other contaminated articles. First aid was administered and those who had suffered high exposures were ordered hospitalized. But additional thousands, many carrying substantial radioactive contamination, managed to flee the affected region through uncontrolled routes, and brought with them contaminated vehicles and clothing. Because the affected region was too large to patrol thoroughly, even by the National Guard, the military frequently had to use force to prevent civilians from breaching the controlled perimeter. In several instances, it was rumored that individuals attempting to re-enter the danger zone against soldiers' orders had been shot.

*   *   *

As a result of the meltdown, thousands of people received fatal radiation doses, many during the evacuation and resulting traffic jams. Some of those who received very high doses died soon after.

Others fell prey to nausea, vomiting, and diarrhea. For some, these symptoms were to abate over a period of weeks. Outwardly, the exposed person would appear to recover. Inside the ends of the victims' bones, however, a kind of marrow death would be occurring. Weeks later, this would usher in the final stage of illness. First the victims' hair would fall out, sometimes completely. Their marrows would no longer vigorously reproduce red blood cells. The people would become anemic and their white blood cell counts would fall drastically, so they would be less able to resist infection. Hemorrhaging of the gums would begin, and body tissues would bruise easily. Ulcers would appear in the mouths, throats, and bowels. Feverish and without appetites, the victims would begin to waste away. Marrow transplants, transfusions, and antibiotics would bring some people through, but would merely unnaturally prolong the agonies of others. Those who would survive the near-lethal doses would often die from acute anemia, cancer, leukemia, or other complications later.

Persons receiving lower doses from the meltdown radioactivity would suffer a variety of conditions depending on the intensity and duration of radiation exposure, as well as on the parts of their bodies where doses were absorbed, and on the person's susceptibility to radiation. Some victims would develop cancers and leukemias years later, others would become sterile. In addition, many unknowingly would suffer genetic injuries, shortening of lifespans, injuries to bone marrow and lymph nodes, and thyroid injuries. Some exposed children would begin growing more slowly and ultimately would show signs of mental retardation.

Environmental effects in the crisis zone reached nightmarish proportions following the meltdown. Waters in the affected zone became undrinkable and were to remain so until each reservoir's natural drainage and refilling processes flushed the tasteless, odorless, and colorless poisons farther downstream from the watershed. Careful tests would be necessary to determine whether radioactivity that accumulated in the bottom sediments of the reservoirs could become resuspended later and recontaminate drinking water above safe limits. The land on which the concentrated radioactive fission products alighted was also thoroughly contaminated with long-lived radioactive elements. The soil would remain unusuable for agriculture until parts of the affected topsoil were carefully scraped off by bulldozers and hauled to a waste storage site, or until the land was cleansed by nature. Winds gradually would scatter some of the resuspendable radioactive particles, driving them elsewhere, and rains would wash other particles from the soil into the area's drainage. While these poisons were dissipating and migrating to other regions, years and possibly many generations might pass before badly contaminated land around Santa Bonita could again be used for cropping.*

* * *

The preceding events, so far, are simply a nuclear nightmare that one hopes will never come true. Moreover, the scenario described — complex as it is — is necessarily a simplified representational accident.[2] In the interest of brevity and clarity, it glosses over many complex processes and incidents needed to precipitate a real nuclear disaster. Yet the Santa Bonita accident sequence and consequences are consistent with currently known facts about the possible course of a major nuclear power plant disaster. And the accident consequences are consistent with official analyses of possible catastrophes made by the Atomic Energy Commission.

The real possibility that a horrendous disaster such as the Santa Bonita accident actually could occur explains why, when the Fermi breeder reactor suffered a partial core melt in 1966, officials seriously considered ordering evacuation of Detroit's one and a half million inhabitants.[3]

Although a "Santa Bonita" meltdown is unlikely at any one plant and most abnormal occurrences at a reactor would not cause a meltdown, the following pages show that the industry-wide chances of a disaster over the next thirty years are much greater than government and industry officials would have us believe. Yet these powerful forces continue to insist that nuclear fission power is safe enough for general use.

Chapter 2 shows how close the southern U. S. has come to a real nuclear catastrophe, possibly like the Santa Bonita meltdown, and it traces the course of a real nuclear power plant accident in 1975, followed by some thought-provoking scientific estimates of accident consequences.

---

*The degree of contamination and the duration of the effects would be proportional to the land's proximity to the reactor and upon the particular isotopes released.

NOTES

1. Henry W. Kendall, Nuclear Power Risks, *A Review of the Report of the American Physical Society's Study Group on Light Water Reactor Safety* (Cambridge, Mass. Union of Concerner Scientists).

3. In a real catastrophe, many on- and off-site support groups would assist control room operators during the emergency, and an Emergency Plan would be put into effect. Your local utility is obliged to prepare such a plan for each of its nuclear power plants, and to file a copy of the plan with the U.S. Nuclear Regulatory Commission. Copies are available to the public in public document reading rooms maintained by the Nuclear Regulatory Commission, or through the utility itself.

3. For a popularly written account of the Fermi accident, see *The Perils of the Peaceful Atom*, by Richard Curtis and Elizabeth Hogan (New York: Doubleday and Co., 1969), or *Nuclear Energy: Its Physics and Social Challenge*, by David Inglis (Reading, Mass.: Addison-Wesley, 1973), or *We Almost Lost Detroit* by John G. Fuller (New York: Readers Digest Press,1975)

Figure 1-1 The 575 MW Connecticut Yankee Nuclear Plant, Haddam Neck, Connecticut

2

REACTOR ABC's

This chapter contains a brief explanation of the physical processes that make nuclear reactors work. Every effort has been made to present this material in everyday language, but some specialized terms have inevitably crept in. A familiarity with the terms and processes will be helpful in understanding the succeeding chapters on nuclear energy.

Nuclear energy, the energy inherent in the nuclei of atoms, is a broad term encompassing fission as well as fusion energy. But fission power is the only commercial nuclear power in existence today, so the following analysis is limited to fission. Fusion, another form of nuclear energy, might revolutionize life on earth someday, if scientific barriers to its use and accompanying environmental problems are solved at a competitive cost. But this probably will not happen until near the end of this century, if at all.

In nuclear fission, tremendous amounts of heat energy are released when uranium-235 atoms under bombardment by atomic particles known as neutrons, absorb a neutron and split into lighter elements such as strontium and iodine.* The splitting of the uranium also releases other neutrons which repeat the process in a

---

*Natural uranium consists of uranium-235 and uranium-238, each having a different atomic mass number designating the total number of protons and neutrons in its nucleus. Of these two similar elements, only the relatively rare uranium-235 is fissioned in a reactor. The uranium-235 comprises less than 1 percent (.71 percent) of natural uranium.

chain reaction. Heavier elements are also created when some of the uranium-238 atoms do not split but are transformed by the absorption of neutrons into elements such as plutonium-239.

Many elements created as a result of fission are unstable, meaning they lose energy by emitting particles. This energy loss is known as radioactivity. These emissions (measured in units like the curie and the rad — see page 47) are dangerous to living things because they can disrupt genes and tissues. Fission power is unique among all modes of energy in that no other energy technology can add as much radiation to natural background levels.

The allure of fission is based on the immense energy that can be released from a tiny volume of highly processed uranium-235. Just one ounce, when fissioned, produces as much heat as burning one hundred tons of coal. Therefore, if enough high-grade uranium ore could be found to process, nuclear power plants would require less mining and cause less water pollution due to mining than would coal burning plants. Black lung disease and typical mining injuries would also be lower in mining high-grade uranium than in underground coal mining.

Because nuclear power plants do not burn hydrocarbons like coal or oil, they do not produce the conventional air pollution that results when the carbon, sulfur and nitrogen of fossil fuels burn. Instead of such harmful substances as sulfur dioxide, nitrous oxides, carbon dioxide and monoxide, ash, and small particulate matter, nuclear power plants in routine operation emit only small amounts of radioactive gas, usually from a stack.[2] In addition, the cooling water effluent routinely discharged by the plant is also slightly radioactive even though it has not been circulated inside the reactor vessel. Against this rather uncritical profile of nuclear power must be weighed the risks and environmental effects of nuclear power discussed elsewhere in this book.

A nuclear fission power plant is basically a machine to produce heat. The heat released during fission is used to turn water to steam which, when directed against the blades of a turbine, creates electricity by rotating a generator — a metal coil that spins within a magnetic field. Thus, once the heat has been generated in the reactor, its transformation to electricity is basically the same as in a conventional power plant in which a fossil fuel (such as coal, oil, or gas) is burned to produce the initial heat.

The fission reactor of a large nuclear plant is a bullet-shaped pressurized chamber more than forty feet high and more than fourteen feet in diameter. Its walls are usually made of steel, at least eight inches thick. About one hundred tons of uranium fuel are suspended in racks within this chamber.

Figure 2-1. Uranium oxide reactor fuel pellets

The fuel is in the form of cylindrical uranium dioxide pellets, about half an inch in length and three-eighths to one-half inch in diameter. The pellets are packed inside twelve-foot-long tubes made of zirconium-alloy metal, which have spaces between them so water can circulate along their lengths. About forty thousand fuel rods, packed into bundles of fifty to two hundred rods each, are used to fuel a large reactor.

The intensity of the fission reaction depends on many factors, including the degree of enrichment of the fuel, its mass, and its physical arrangement in a given volume. When fuel is properly loaded into a reactor, its geometric arrangement permits a limited, carefully controlled amount of fission to occur.

Normally, the fuel load is surrounded by water under enor-

Samuel A. Musgrave, Atomic Industrial Forum

Figure 2-2 Bundles of uranium fuel elements

mous pressure: about 2,250 pounds per square inch in a *pressurized water reactor* and 1,000 pounds per square inch in a *boiling water reactor.* The pressurized water reactor and the boiling water reactor are the two main kinds of reactors in use today. Water in their cores both cools the fuel and slows down (or moderates) neutrons released by the uranium during fission. The neutrons must be slowed in order to sustain the chain reaction, which is necessary for the plant to produce electricity.

The energy of the chain reaction in the reactor core heats water in the reactor vessel surrounding the core. In the pressurized water reactor, the water is kept under such high pressure that it cannot boil athough heated to 600°F. From the reactor vessel (also called the reactor chamber), this highly radioactive water is piped to a steam generator.

At the steam generator, the water flows through thousands of thin tubes and their large surface area facilitates the heating of another completely separate stream of water, kept at lower pressure, and turns it to steam. (The lower the pressure of water, the easier it boils and turns to steam.) This steam from the steam generators then spins plant turbines and turns an electrical generator. After running the turbine, the working steam is returned to a liquid state again in a condenser and is piped back to the steam

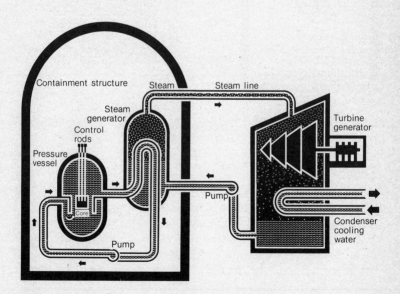

Figure 2-3. Pressurized Water Reactor (PWR)

generator. Meanwhile, the first or primary stream of extremely hot pressurized water directly from the reactor vessel is returned there for reheating.

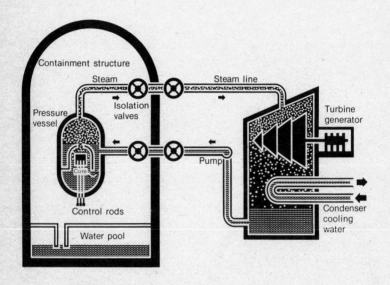

Figure 2-4 Boiling Water Reactor (BWR). Steam passes *directly* from the reactor through the turbine, which also must be shielded.

In the boiling water reactor, because its pressure is lower than in the pressurized water reactor, water boils in the reactor vessel (at 545° F.) and highly radioactive steam results. From the top of the vessel where the steam has risen, the vapor is piped to the plant's steam turbines. In contrast to the pressurized water reactor, no "secondary stream" of isolated working steam circulates in the boiling water reactor. The turbines are turned by the radioactive water directly from the reactor core. The used steam is then sent to a condenser and is returned as water to the reactor vessel.

To control the fission process in either a pressurized water or boiling water reactor, long rods made of neutron-absorbing boron or cadmium are inserted into the reactor between the fuel bundles. These can be raised or lowered in the reactor to regulate the rate of fission and the power output of the reactor.

During reactor operation, water must circulate within the reactor core to remove heat from the fissioning fuel, which reaches 4,300°F at its center. The water must keep on circulating even after control rods have been placed in their shutdown position to stop the nuclear chain reaction. Even after controlled uranium fission has been stopped the fission products (such as strontium and iodine, which have accumulated in the core, continue their radioactive decay unabated. This residual radioactivity can generate enough heat to melt the core anytime the flow of cooling water stops. A major primary cooling system rupture due to a pipe break would interrupt the coolant flow, causing the superheated pressurized cooling water to escape at a great rate, leaving the core dangerously uncooled.

In that event, a back-up cooling system powered by high pressure nitrogen gas is designed to flood the core and pressure vessel with cool borated water. If this device fails to provide the necessary cooling, then the Emergency Core Cooling System would be required to reflood the reactor with more water. If it failed to do so for any reason, the reactor would begin heating up at a tremendous rate. Although the loss of water would have brought controlled fissioning to a stop, the heat caused by the fission products accumulated in the uncooled reactor would cause the fuel rods to heat up rapidly to more than 5,000°F — at the rate of 400°F every ten seconds. Within less than a minute, the fuel rods and surrounding metals would have melted and collected in a glob of molten steel, uranium, and assorted radioactive poisons at the bottom of the reactor vessel. In about an hour, the fiery glowing material would melt its way through the reactor vessel. Chemical explosions might also occur, hurling metal against the containment building. Radioactive gases would generate strong, possibly explosive pressures against the containment shell. Beneath the reactor vessel, no man-made structure could contain the fiery liquefied metals so they would melt through the concrete floor of the plant, reacting with the chemicals in the concrete. The incandescent mass would eventually cool enough to resolidify, but not before some portion of the reactor's radioactivity escaped the containment shield and was lofted into the air by wind or mingled with ground waters.

The seriousness of the nuclear accident's consequences would depend, among other things, on whether the reactor had just been fueled, in which case the fuel rods' radioactive burden would be slight, or whether they had been fissioning for a year and hence were at their maximum toxicity.

The core of a large fission reactor after sustained operation contains as much long-lived radioactivity as could be released by

Figure 2-5. 800 Ton Nuclear Reactor vessel, Commonwealth Edison's Dresden-2 Unit

Babcock & Wilcox

the explosion of one thousand Hiroshima-type atomic bombs, although the kind of radioactivity that would be released in a meltdown would differ from the kinds of elements released in a bomb. Because of the enormous quantities of radioactive materials in a reactor, the radioactivity must be kept almost perfectly contained under normal operating conditions, even in routine plant emissions, through stacks and waste water. An accident that caused a radioactive discharge to the environment equal to even a few percent of a fully loaded core's burden could be a disaster of gigantic proportions.

Nuclear industry representatives frequently stress that nuclear reactors cannot explode "like a bomb," but they do not point out the respects in which a reactor meltdown could be worse than a bomb blast. The radioactive fuel load of a reactor is larger than a bomb and more atoms are ultimately split in the reactor than in even a large nuclear blast. Whereas the blast lasts for only a fraction of a second, fissioning in the reactor occurs constantly during reactor operation. The result is a radioactive inventory larger than that created by a large atomic bomb. Moreover, when a bomb explodes, its fission products are blown up into the atmosphere, whereas when a reactor releases fission products as a result of a meltdown, it doesn't have the same explosive force, so fission products are released horizontally at lower elevations where they can do more damage.

Physicist John P. Holdren of the University of California has calculated that a large plant's radioactive inventory consists of "about 12 billion curies of fission products, 3 billion curies of actinides, and 10 million curies of activation products."[3] Professor Holdren has concluded that half the amount of just one of these radioactive products, strontium-90, would be enough to contaminate all the fresh surface water in the forty-eight adjoining states to six times the maximum permissible concentration for the poison, according to federal standards. And strontium-90 remains dangerous for many generations. Holdren also found that a quarter of the radioactive iodine-131 in the large reactor would suffice to contaminate the atmosphere over the same area to a height of about six miles. It is by no means likely that the radioactivity would be distributed in this manner, but the presence of such quantities of radioactivity suggests the potentially catastrophic nature of nuclear power plant accidents.

Figure 2-6. Liquid-Metal Fast-Breeder Reactor (LMFBR)

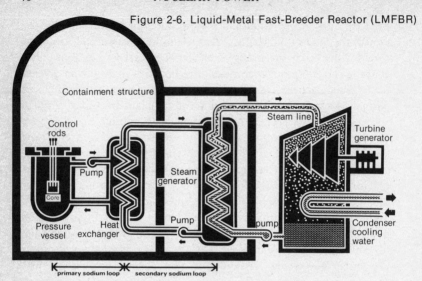

Most government energy research funds are spent on the development of a "fast" fission reactor, the *liquid-metal fast-breeder reactor* (LMFBR). The "breeder" reactor is intended to do two things — generate heat for steam and at the same time produce more reactor material than is consumed. The core of the "breeder" uses plutonium-239 as a fissile material which breaks up into smaller fragments, with the release of energy and additional neutrons, just like uranium-235. Unlike uranium-235 plutonium can be made to fission with fast neutrons. Surrounding the core is a "blanket" of uranium-238 — the common isotope — and the greatest quantity of neutrons given out by the core are absorbed by the nucleus of the urnaium-238 atoms, transorming them into plutonium-239. If operated hard enough, more plutonium is made in the blanket than is used up in the core, hence — "breeder." Water cannot be used as a cooling medium because it slows down the fast neutrons, therefore the scheme to use liquid sodium for cooling. The liquid sodium is then passed through a steam generator, and the steam from the output side of the generator drives a turbine.

An important factor in LMFBRs is the "doubling time" — the time required to double the amount of plutonium in it. Further details will be found in Chapter 7.

Just how securely these radioactive inventories are being held can be surmised from Chapter 3, in which we will consider a real reactor accident in 1975 at a large U. S. nuclear power plant.

NOTES

1. For a discussion of radon dangers, see Arell S. Schurgin and Thomas C. Hollocher, "Lung Cancer Among Uranium Mine Workers," *The Nuclear Fuel Cycle* by D. F. Ford, *et al* (San Francisco: Friends of the Earth, 1974).

2. These "routine emissions" will be discussed further in Chapter 5.

3. "Actinides" are those elements listed after Actinium on the Periodic Table of Elements. "Activation products" are elements made radioactive by neutron actions. Such produces may include a wide variety of radioactive elements.

Radioactivity is the property of certain elements to disintegrate. As they do, they may give off dangerous forms of energy. The energy may be in alpha particles (two protons and two neutrons), beta particles (equivalent to an electron or positron), or gamma rays (a high frequency, high energy emission). Whereas gamma rays can penetrate thick barriers, alpha particles can be stopped by a sheet of tissue paper. Beta particles have intermediate penetrating power.

Radioactivity is usually measured in curies or rads. Any radioactive substance that is disintegrating at the rate of 37 billion atomic disintegrations per second is said to be emitting one curie of radiation. This value was chosen because it happens to be the number of atomic disintegrations emitted every second by a gram of radium.

Whereas a curie is a measure of radiation emitted, a rad is a measure of radiation absorption. When an organism receives a radiation dose causing 100 ergs of energy to be deposited per gram of tissue, we say that one rad had been administered. One rem is simply a rad, multiplied by a number that adjusts its value to reflect the biological potency of the particular kind of radiation received, be it alpha, beta, or gamma emissions. For most purposes in this book, a rad and a rem are functionally equivalent.

Figure 3-1. Control room of the Rancho Seco Nuclear Power Plant

3

RISKS WE SHOULDN'T TAKE

The control room of a reactor is both a showcase of electronic wizardry and the temple altar of a nuclear religion. That religion places ultimate faith in the power of science to tap energy from the cores of radioactive atoms, without causing human harm from their deadly radioactivity.

The control room is usually calm, immaculate, and deceptively neat. The tidiness masks the phantasmagorical complexity of the computerized nuclear instrumentation, connected by thousands of sturdy cables that funnel millions of electronic signals into the control room.

These signals are neatly displayed on a panorama of gauges, dials, and indicator lights surrounding the control room operator. The ever-changing masses of data come from myriads of sensors that take the pulse, heartbeat, and temperatures of a gargantuan nuclear reactor, seething quietly like a steel dragon in a cave of reinforced concrete nearby.

Although the apparent omnipotence of science pervades the control operator's domain, the room has a theatrical aura, too, and conjures up visions of futuristic Hollywood Captain Marvel adventures. Yet it also suggests the sober efficiency of an Apollo launch center.

From the inner sanctum of the plant's control room, the plant machinery that regulates the immense, primeval energies of the atom is itself regulated. Without this fine tuning, an efficient, smoothly operating reactor could rapidly metamorphose into an uncontrollable raging furnace.

Near Athens, Alabama, the Tennessee Valley Authority operated two large nuclear reactors at its 2.2-billion-watt Brown's Ferry Nuclear Power Plant. But the plant had to close in March 1975 after a grim accident with nationwide repercussions. As of February 1976 the plant remains shut.

The two Brown's Ferry reactors, Unit 1 and Unit 2, have one common control room. Beneath it is a cable-spreading room where cables governing Unit 1 are separated from Unit 2 cables and are routed through tunnels into their respective containment buildings. Virtually everything electrical in the plant, including routine and emergency reactor cooling systems, is controlled by cables that pass through the spreading room. With so much cable side by side, an accident in the cable room can disable both reactors simultaneously. The chances of a nuclear meltdown would be increased if the accident required control room evacuation, leaving the reactors without human guidance.

As mentioned in Chapter 2, the colossal amounts of heat generated in reactors by the splitting of uranium fuel is kept carefully regulated by large amounts of cooling water and by special control rods. Without its cooling water, a reactor core gets so intensely hot that, *even if shut down*, it continues to get vastly hotter until it melts. This can lead to a major catastrophic discharge of radioactivity to the environment.

Larry Hargett, an electrician for the Tennessee Valley Authority was testing for an air leak in the cable-spreading room of the Brown's Ferry plant.[1] Air leaks are important to detect, because dangerous radioactivity can escape from nuclear reactor buildings if they are not sufficiently airtight, especially in the event of an accident.

Hargett and another man were stuffing flammable polyurethane foam strips into spaces around the plant's electrical cables leading from the control room to the two reactors. After stuffing one hole with foam, the men tested it, according to usual plant procedure, by holding a lighted candle with an open flame close to the hole. But the hole they were testing had not been completely sealed. The flame of the candle was sucked horizontally into the opening.

This fusing of medieval and space-age technology ignited some foam. The contents of chemical fire extinguishers that the electricians sprayed on the fire were drawn ineffectually through the hole in the wall by strong drafts. Fifteen minutes later, the electricians told a guard about the fire. After more fumbling, a fire alarm was finally sounded.

Meanwhile, bizarre phenomena had begun to occur in the control room. The fire, which ultimately was to damage sixteen hundred cables, was already causing electrical chaos on the control board. Lights were going on and off erratically, alarms were sounding, smoke was coming out of one panel, and in the plant, equipment was disobeying control signals.

After ten minutes of this kaleidoscopic display, the operator attempted to shut down reactor Unit 1. Although critical reactor control systems rapidly were becoming inoperable as their cables melted, the operator managed to insert Unit 1's control rods, bringing the atom-splitting process in the reactor to a halt.

The danger of a nuclear disaster that would spread radiation was lessened but by no means eliminated. In fact, an intense struggle to prevent the reactor core from melting was only beginning.

The operator now found that not only the reactor core's normal cooling system, but its emergency system, with all five subsystems, had been electrically disabled, along with key instruments relaying critical reactor temperatures and pressures to the control room. It was a mere forty minutes or so after the onset of the fire, but the only water supply the reactor now had was from the control-rod drive pump, a small pump never designed to be the primary source of reactor cooling water.

The efforts to control Unit 1 were hampered rather than aided by the plant's own firefighting equipment and procedures. When the cable room's carbon dioxide fire protection system was activated, it malfunctioned, driving thick smoke and fumes into the control room.

As the room began filling up with smoke, engineers began coughing and choking. With evacuation of the control room imminent, the CO_2 system was turned off. If abandonment of the controls had been necessary, the Brown's Ferry plant, with its power-less computer and scrambled electronic brains, would have been essentially on its own.

Forty-five minutes after the fire started, all of the instrumentation for reactor Unit 1 failed. But even without instruments, the operators knew the reactor could not safely depend on the control-rod drive pump for cooling. During the battle to cool Unit 1, more than three quarters of the water normally above the reactor core was lost. The top of the core came within just forty-eight inches of exposure to the air.

The Brown's Ferry fire was only marginally kinder to reactor Unit 2 than to Unit 1. Unit 2's entire high-pressure emergency core cooling system was disabled, and so were some of its low pressure systems. Only a jerry-rigged cooling hook-up brought Unit 2 under control nearly two hours after the fire started. Normal shutdown could not be achieved on Unit 1 until almost fourteen hours later.

Not only did the fire cause the sequential, interrelated failures of plant mechanical systems, it uncovered numerous failures by personnel to observe correct safety and emergency procedures. Most serious of all, the fire disproved the claim by nuclear proponents that the "redundancy" of reactor safety features makes meltdown well-nigh impossible. If one fire in the cable room can knock out all emergency devices, then the nuclear proponents' assurances of plant safety are unfounded. Every nuclear power plant in the U.S. is designed with a cable room that is vulnerable to accidents.

The estimated $100 million cost of the Brown's Ferry fire would be a small price to pay if the public learns to distrust the nuclear advocates' safety claims. It is a small price, too, compared to the billions it would cost to immediately close all nuclear plants in the U.S. and install more reliable fixed firefighting equipment in them and in all nuclear plants now under construction. This latter course would be extremely unpopular with utilities and their investors, and it would certainly cause a national crisis of confidence in nuclear power as a reliable power source.

At Brown's Ferry, because a meltdown was averted, albeit narrowly, no accidental release of radioactivity occurred. If it had, how serious might the consequences have been relative to those in the Santa Bonita scenario? To find out, we will review some recent landmark studies on this issue.

OFFICIAL CATASTROPHE ESTIMATES

In the 1950s when nuclear reactors were small and few in number, the Atomic Energy Commission called upon the Brookhaven National Laboratory for an assessment of the possible consequences of a major nuclear accident. The result was the Brookhaven Report (WASH-740),* which has been the bane of the AEC ever since its publication in 1957. The report estimated that a maximum credible reactor accident (that is, the largest the AEC deemed plausible) could cause 3,400 deaths and $7 billion in property damage. This report assumed that the affected reactor was small (100-200 MW),† and that 50 percent of the reactor core's radioactivity escaped.

The estimates of the Brookhaven Report in some respects were conservative. The report assumed that an accident occurred in a reactor thirty miles from a city. Reactors are now being built twenty-four miles from New York City; twelve miles from downtown Gary, Indiana; ten miles from Philadelphia; five miles from Trenton, New Jersey; and four miles from New London, Connecticut.

Theoretical Possibilities and Consequences of Major Accidents in Large Nuclear Power Plants. United States Atomic Energy Commission (WASH-740), March 1957.
†The symbol Mw stands for megawatt, which means "a million watts." A watt is the power required to lift 2.2 pounds four inches per second.

The report's death toll projections do not include deaths from genetic defects and latent cancers.

To see how the Brookhaven Report results might apply to a large nuclear power plant today, the Committee for Nuclear Responsibility in Dublin, California, adjusted the computations so they yielded data applicable to a 1,000 Mw reactor subject to meltdown thirty miles from a city of a million.* Unlike the Brookhaven Report, however, the Committee for Nuclear Responsibility assumed that only 10 percent (and not 50 percent) of the core's radioactivity was released.

These are the conclusions that emerged:
— Agriculture and water supplies would be ruined in an area larger than California.
— 3,340 people would die up to a hundred miles away from acute radiation exposure.
— Assuming 600,000 people were exposed to 100 rads, or 1,200,000 each to 50 rads, or 2,400,000 people each to 25 rads, then 50,000 people would die of cancer, including leukemia, years later. Uncounted birth defects, stillbirths, and mental retardation would occur in offspring from irradiated parents.
— Under certain circumstances, for example, if evacuation of a major city was stalled by a monumental traffic jam, deaths could skyrocket above these projections.

Furthermore, the AEC itself in 1964 and 1965 reappraised the consequences of a major nuclear accident and found the results so chilling it did not release its findings until 1973, after strong pressure from environmental groups.† The AEC updating indicated that the worst possible accident could kill 45,000 people, injure 100,000, and do $17 billion damage (in 1965 dollars). Land-use restrictions might persist for five hundred years downwind of the accident throughout an area the size of Pennsylvania. Poisonous fission products like strontium-90 and cesium-137 — two radioactive elements also found in nuclear weapons fallout — would be deposited.

---

*The committee is a nonprofit organization opposed to nuclear power and is directed by a board of eminent scientists and public figures, including four Nobel laureates.

†Attorney Myron Cherry obtained the document by threatening to sue the AEC under the U.S. Freedom of Information Act. Cherry participated in AEC hearings on emergency safety systems as an attorney for the Consolidated National Intervenors, a coalition of environmental groups opposed to the licensing of nuclear power plants.

They can contaminate air, water, soil, plants, and animals. Through a biological concentration process, those plants and animals would effectively magnify the strontium and cesium radioactivity.*

Foreboding as the Brookhaven Report and its suppressed update were, neither document was as ominous as a far less well known study which was marked classified and squirreled away in the AEC's bureaucratic labyrinth almost as soon as it was completed. Prepared by the Engineering Research Institute of the University of Michigan at Ann Arbor and dated July 1957, the report is called, *"Possible Effects on the Surrounding Population of an Assumed Release of Fission Products into the Atmosphere from a 300 Mw Nuclear Reactor Located at Lagoona Beach, Michigan."* (Lagoona Beach is the site of the Detroit Edison Company's ill-fated Fermi breeder reactor, which suffered a core melt accident.) The report projected that under maximum radioactive release conditions, 133,000 people would receive a lethal dose of 450 rads and half would perish, instead of the 3,400 immediate fatalities estimated in the Brookhaven Report. And while the Brookhaven Report set the injury toll at 43,000, the University of Michigan report which the AEC quietly buried stated that as many as 181,000 people could receive 150 rads in a catastrophic accident. A significant number of these injuries would prove fatal.

THE CHANCES OF DISASTER

How might a catastrophic accident occur, and what are its chances? The AEC has repeatedly assured the public that the chances of a disaster at any one particular plant are extremely small because of the presence of elaborately engineered safeguards like the Emergency Core Cooling System and the multi-storied, steel-clad, reactor containment building. But will the ECCS work? The Zion Station nuclear plant north of Chicago, which opened in 1973, operated into 1974 with its ECCS incorrectly wired.[2] But flawless ECCS operation depends on more than correct installation by fallible humans.

---

*For instance, the flesh of a fish grown in fresh water containing cesium-137 will have 1000 times as much cesium-137 as the water in which it was grown. Milk from cows grazed on grass laced with strontium-90 will have a more concentrated strontium-90 content than the grass. The concentrated radioactivity will then be passed on to anyone drinking the milk.

Many engineers and scientists, including some at AEC headquarters and in AEC laboratories who themselves worked on the ECCS, have expressed serious reservations about ECCS engineering reliability.[3] An internal memorandum written in February 1971 by Dr. Milton Shaw, then director of the AEC's Division of Reactor Development and Technology, conceded that "no assurance is yet available that emergency coolant can be delivered at the rates [of flow] intended and in the time period prior to . . . fuel melting due to decay heat generation."[4] Dr. Shaw, incidentally, was a prominent nuclear advocate who publicly defended AEC nuclear safety programs.

ECCS safety test results have been both fragmentary and disturbing. Although individual ECCS components have been tested, so far the whole system has only been tested by controversial computer simulation and in a "miniscale" model. When six of these miniscale ECCS tests were administered to the system in the fall of 1970 for the AEC by Aerojet Nuclear Company in Idaho, the ECCS failed the tests all six times. In one episode, when a cooling line on the test model was deliberately broken, the emergency coolant was blown out of the break instead of flooding the miniature core, which was only nine inches in diameter. Although the tests were too small to establish convincingly the safety of the ECCS, had they been successful, their failure has seriously undermined public confidence in reactor safety systems. The model ECCS clearly did not work as the computer had predicted.

New tests were scheduled for 1974, but their completion has been postponed to 1976-1977. The ECCS issue remains very much unsettled, and engineers at the Idaho National Engineering Laboratory in Idaho Falls are now constructing a small reactor on which to conduct semiscale tests. No full-scale tests are planned.[5]

A number of articulate, vocal, and effective opponents of nuclear power were aroused by the AEC's apparent failure to demonstrate the safety of the ECCS before approving its installation on U.S. reactors.

The Union of Concerned Scientists of Cambridge, Massachusetts, took a leading role in a protracted battle with the AEC over the ECCS. Led by Daniel F. Ford, an economics graduate from Harvard, and Professor Henry Kendall, a professor of physics at Massachusetts Institute of Technology, UCS conducted intensive, detailed technical assessments both of reactor safety systems and of the AEC-sponsored tests designed to establish their reliability. In national public hearings on the ECCS design, held during 1972-1973, the UCS — serving as the technical adviser for the Consolidated National Intervenors* — challenged the premises of the AEC's safety program and the rules which the AEC proposed to use in accepting ECCSs in nuclear plants. Focusing on the lack of assured safeguards for the prevention of a meltdown in the plants, UCS said that the AEC's reactor safety claims are based on an experimental and analytical structure that could not even withstand a casual examination:

> It is not simply that one or two vital links are demonstrably weak, or indeed missing. It is that the whole structure is unsound. It is based on engineering results whose quality and scope are broadly and seriously deficient and whose interpretation is in serious dispute by a host of persons whose views cannot be disregarded. The structure depends critically on computational predictions that are critically weak and unreliable, and which are themselves in serious dispute.

Recent events have tended to confirm the UCS analysis. During the ECCS hearings, the AEC had conceded that a steam generator rupture could disable the ECCS. But the commission had contended that such an accident was so unlikely it could be discounted. No such rupture ever occurred. Then on February 26, 1975, steam generator piping ruptured at Wisconsin Electric's Point Beach plant at Lake Michigan, forcing the plant to close and dispelling yet another AEC myth.[6]

In concluding its ECCS challenge, UCS asserted that the licensing of nuclear reactors was unjustified because of "inadequate experimental understanding of LOCA [loss of coolant] phenomena [and] inadequate experimental confirmation of ECCS capability," as well as for other reasons.[7] The AEC's response to the UCS challenge was to change its ECCS design standards slightly. But the AEC's credibility was mortally wounded and the confrontation over the ECCS no doubt hastened the AEC's demise.

---

*A coalition of environmental groups fighting against the nuclear reactor program on safety issues.

†On January 19, 1975, the new U.S. Energy Research and Development Administration took over all the functions of the AEC except for licensing and regulating nuclear power plants. These were vested in a new U.S. Nuclear Regulatory Commission.

The ECCS battle helped stoke a national controversy over nuclear power plant safety. One salient issue was the AEC's inconsistent approach to assessing the chances of loss-of-coolant accidents. For example, the failure of the large, pressurized steel reactor vessel has traditionally been regarded by AEC safety researchers as virtually impossible.[8]* Yet the pressure vessel is designed to the same standards as major reactor pipes, and pipe ruptures *are* regarded as possible. Peter Morris, a member of the Directorate of AEC Regulatory Operations, said in 1973:

> . . . [Within] the AEC it has been the policy that designs should not be required to provide protection against pressure vessel failure. So the question of whether or not such an event was credible did not arise. The reason is very simple — no design was available for a building which could withstand the consequences of pressure vessel failure, so it was decided to accept the risk.[9]

Policies like these have led the AEC to rely on inadequate safety measures.

With reactor safety technology under assault from many fronts and the credibility of the ECCS badly damaged, the AEC's successor, the Energy Research and Development Administration, began searching for a new safety panacea. The latest "solutions" under consideration are the building of nuclear power plants underground or in big nuclear energy complexes euphemistically called "nuclear energy parks." Neither of these plans is satisfactory, as the reader will see later. In response to the AEC's public relations program about reactor safety, at least one AEC safety researcher has opted out.

Reactor safety expert Carl J. Hocevar was once responsible for developing a basic method used by the AEC to analyze fission plant safety. In resigning from the AEC in September, 1974 to work for the Union of Concerned Scientists, Hocevar stated:

> [Despite] the soothing reassurances that the AEC gives to the uninformed, misled public, unresolved questions about nuclear power safety are so grave that the United States should consider a complete halt to nuclear power plant construction while we see if these serious questions can, somehow, be resolved.[10]

---

*The Energy Research and Development Administration.

Coincidentally, the same day Hocevar's resignation was publicized, the AEC was obliged to order sixty-day shutdowns for twenty one of the fifty operating nuclear plants in the U.S. to check for leaky pipes. Leaks of radioactive primary cooling water had been discovered in two plants, and cracks which had not yet fully opened were found in a third plant.

Many safety problems with nuclear reactors can be correlated with the extremely rapid expansion of the technology. The nation's first commercial reactor opened in 1957; a decade later, only nine were in existence. But eight years after that, by 1975, fifty-six had been built. And most reactors built in the late fifties and the sixties were relatively small. Modern plants generate a billion watts (1,000 MW) of electricity. These behemoths are basically just scaled-up versions of their predecessors. The first 1,000 MW plant only went into service in 1973 and now more than ten are already in operation, with more under construction or scheduled. But the experience base for predicting and preventing accidents with these plants is necessarily far too small to make statistically valid accident predictions. If we have fifteen years of accident-free reactor operating experience with these large plants, we might infer that a major accident is not highly likely within a decade. But the fifteen-year experience base would not be enough to conclude, as the AEC and Nuclear Regulatory Commission did, that a large accident was impossible in twenty years, a hundred years, or a million years.

The dangers of widespread nuclear operations with a limited experience base have been aggravated by the AEC's failure to levy sufficient penalties on nuclear utilities found in violation of federal radiation regulations.[11] Generally, even for serious violations that presented health threats to workers or the public, fines averaged about $6,000 in fiscal year 1974. Compared to the costs of a large reactor shutdown, which can exceed $250,000 per day, such low fines are scant deterrents to large corporations. While the AEC was providing lax safety enforcement and was energetically promoting fission power plants through forceful public relations campaigns (see Chapter 8), its basic safety research effort was limping along.* Critically important Loss of Fluid Tests (LOFT) intended to docu-

---

*Even in fiscal 1976, the Nuclear Regulatory Commission's reactor safety budget of $62 million was termed relatively small by an American Physical Society group that reviewed reactor safety.

ment the viability of the Emergency Core Cooling System were allowed, according to science writer Robert Gillette, to take "more time from start to finish than the Apollo program took to land two men on the moon." First the project was poorly managed, then it hit budgetary restrictions in 1971.[12] A shortage of safety funds led in the same year to the suspension of tests at the AEC's Pacific Northwest Laboratory on how reactors respond to losses of coolant. Milton Shaw, then Director of Reactor Development and Technology for the AEC, described the experiment as "producing rather important safety information."[13] Also in 1971, a vital fuel-rod-test project that was essential to an understanding of the ECCS was canceled at the AEC's Oak Ridge National Laboratory just as its results were substantiating charges by the Union of Concerned Scientists about ECCS design deficiencies. Shortly before their cancellation, the tests showed fuel rods swelling, buckling, and rupturing at temperatures hundreds of degrees lower than those they were supposed to withstand in the event of a loss of coolant.[14] Such failures to meet design standards inevitably cast doubts upon the quality of other reactor components.

In case after case, the AEC sanctioned the licensing and construction of commercial reactors with unproven safety systems at sites where they could endanger millions of people. And while it was defaulting on its regulatory functions and was exposing the public to nuclear accident risks, the AEC was simultaneously trying to shield the public from knowledge of just how bad those accidents might actually be.* By this alliance of irresponsibility and deceit, the AEC nurtured the widespread growth of commercial nuclear power in the U.S.

---

*An exceptionally fine account of how the AEC and the nuclear industry tried to make the results of the Brookhaven revision sound less terrifying, and failing that, how the AEC suppressed the revision, can be found in John G. Fuller's *We Almost Lost Detroit,* cited in Chapter 1

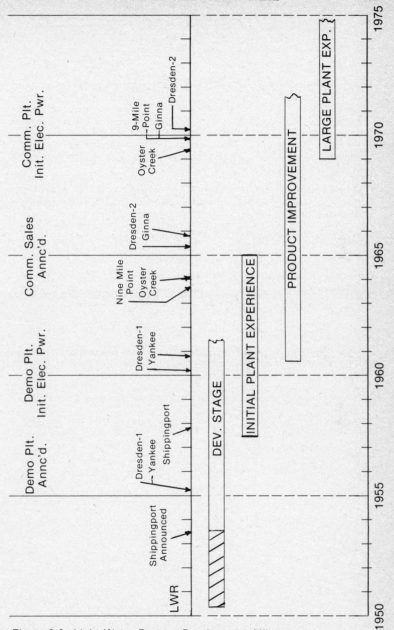

Figure 3-2. Light Water Reactor Development Milestones.

NOTES

1. *New York Times,* March 25, 1975. *Wall Street Journal,* July 31, 1975.

2. *BPI Newsletter,* Vol. 2, No. 1 (Spring, 1975). This newsletter is published by Business and Professional People in the Public Interest, 109 North Dearborn Street, Suite 1001, Chicago, Illinois 60602.

3. Joel Primack and Frank von Hippel, *Advice and Dissent, Scientists in the Political Arena* (New York: Basic Books, Inc., 1974).

4. Quoted in Daniel F. Ford and Henry W. Kendall, *An Assessment of the Emergency Core Cooling Systems Rulemaking Hearings* (San Francisco: Friends of the Earth, 1975).

5. *Wall Street Journal,* July 9, 1975. *BPI Newsletter, loc. cit.*

6. Ibid.

7. Daniel F. Ford and Henry W. Kendall, op. cit.

8. Robert Gillette, "Nuclear Safety: AEC Report Makes the Best of It," *Science,* Vol. 179, January 26, 1973.

9. Julich Meeting, International Atomic Energy Agency, February 5-9, 1973.

10. "Power Reactors face Safety Test," *New York Times,* September 22, 1974.

11. "AEC Penalizes Few Nuclear Facilities Despite Thousands of Safety Violations," *New York Times* (August 25, 1974).

12. Robert Gillette, "Nuclear Reactor Safety: A New Dilemma for the AEC," *Science,* Vol. 173, No. 3992 (July 9, 1974).

13. Quoted in Gillette, *op. cit.*

14. Quoted in Primack et al. *op. cit.*

4

"IMPOSSIBLE" ACCIDENTS

"The probability of occurrence of publicly hazardous accidents in nuclear power reactor plants is exceedingly low. . . . One fact must be stated at the outset: no one knows now or will ever know *the exact magnitude of this low probability of a publicly hazardous reactor accident."* [emphasis added]

> *Theoretical Possibilities and Consequences of Major Accidents in Large Nuclear Power Plants* WASH-740. A report by the Brookhaven National Laboratory for the United States Atomic Energy Commission, March 1957.

History is often made by the occurrence of exceedingly improbable events. Sometimes these may result from the failure of a supposedly fail-safe technology. Two out of seventeen Apollo missions, for instance, failed from human error. And with all the safety precautions of the space program three astronauts were still incinerated on their launching pad. In complex modern engineering systems, one bad weld, one defective valve, one faulty wire can trigger an "impossible" accident. If some of these accidents are catastrophes of enormous scale, society may decide it cannot afford to risk them *at all*, no matter how small their highly disputed chances of occurrence are said to be. To date, not even the nation's largest insurance companies have been willing to take this bet on nuclear power: full coverage for the economic effects of a nuclear catastrophe staked against the odds that a deluge of radioactivity will never spew from a nuclear power plant. (See Chapter 7 for a more detailed discussion of nuclear insurance.)

To counter its critics on the nuclear safety issue, the AEC attempted to base its safety pronouncements on scientifically derived mathematical assessments of nuclear accident risks. The most recent effort of this kind was the Rasmussen Reactor Safety Study (RSS), a two-year, $3-million study on nuclear accident risks and consequences,[1] directed by Professor Norman C. Rasmussen of the Massachusetts Institute of Technology.

Long before the RSS finally appeared in draft form in August 1974, government officials broadcast reassuring conclusions about reactor safety which they derived from the study, without issuing the draft so those conclusions could be critically analyzed. As soon as the draft appeared, a barrage of criticism was leveled at it by skeptical scientists in the environmental movement and by the U.S. Environmental Protection Agency. The RSS draft consisted of a main report and about a dozen additional volumes of technical appendices. These thousands of pages were intended to put to rest once and for all the public fears of nuclear accidents that had been accentuated by the publication of the original Brookhaven report, and the belated appearance of its suppressed update. (Had more people known about the University of Michigan study concerning the Lagoona Beach reactor, that, too, would have aroused great concern.)

Although the authors of the original Brookhaven report felt that a meaningful scientific computation of nuclear catastrophe probabilities could never be made, the Rasmussen study now claimed to have done the impossible, using computers and modern statistical methods. This suited the AEC's purposes to a tee because if the public could be sold on the idea that the chances of a disaster were minute enough, people would cease worrying about how dreadful the accidents might be, and the impact of the earlier AEC accident studies would be diluted. Unfortunately for the nuclear establishment, prestigious critics assailed basic premises of the study and weakened its credibility. The Rasmussen group responded with a number of modifications and the new Nuclear Regulatory Commission released the final report under its imprimatur in 1975.

Although this final document made concessions to critics by acknowledging that major accident consequences could be more severe than the draft stated ($8 billion more severe in the catastrophic case), the final report hewed closely to the study group's initial findings that the odds of a grave nuclear power plant accident are infinitesimally small. Because the Rasmussen study is the U.S. atomic community's current basic defense of fission reactors, it is worth analyzing in detail.

The study attempted to assess all the possible events that might lead to core melts of varying severity, and to assess the likelihood of each event. By estimating in detail the chances for specific malfunctions and tallying them, the Rasmussen study group hoped to determine accurately the chances of every possible significant accident. The RSS considered core melts with relatively minor consequences and those with catastrophic consequences.

Comparing the risks of nuclear accidents to various known risks (like car fatalities and tornadoes), the study concluded that the chance of one person dying from a nuclear accident was a scant one in 5 billion per person per year — roughly 166,666 times less than from all other possible accidents on earth. And the chances of 1,000 people dying annually from a nuclear accident were found to be about the same as 1,000 people dying from meteors. Not too many "shooting stars" penetrate the earth's atmosphere and fewer still hurt anyone.

The study also found that the more severe the accident studied, the smaller were the chances of its happening. The chance of a melt with the worst conceivable consequences was judged to be one in a billion per reactor per year. For the most likely and least severe kind of core melt, with 100 reactors assumed operating, the chance of the core melt is said to be one in 200.

The health consequences of the most likely core melt — one caused by a small pipe break — were rated as minor. Essentially no deaths, no injuries, and no genetic consequences were anticipated by the Rasmussen group for that event. Property damage from this less serious accident was estimated at less than a million dollars. (This excludes damage to the reactor and the hundreds of millions [or more] it might cost the operating utility to supply power from an alternate source and repair or replace the reactor.)

The consequences detailed by the RSS for the worst conceivable core melt were grave indeed, but remarkably less severe than previous AEC studies in 1957 and 1965 had found. The new maximum consequences included 3,300 fatalities, 45,000 "early illnesses," and more than 1,500 latent fatal cancers, which might appear many years after the meltdown. A 290 square-mile area would have to be evacuated and property damage would exceed $14 billion. There would also be latent genetic damage roughly equal to the number of latent cancers.

The RSS study also assayed the economic losses in the U.S. from various causes such as automobile accidents, fires, and hurricanes and found "reactor accidents have a negligible impact on the total risk of economic loss from man's activities and from natural events."

While still in its draft form, the RSS's numerical conclusions were substantively challenged by several authorities.[2] The U.S. Environmental Protection Agency, for example, said the draft's estimates on deaths and injuries resulting from nuclear accidents were about ten times too low. The Union of Concerned Scientists found the draft casualty figures sixteen times too low.[3]

Physicist Amory B. Lovins, a British representative of Friends of the Earth, concluded that the RSS results are simply invalid. In a devastating critique attacking virtually everything in the RSS, Lovins cites impressive scientific references to document his charges.[4] Although Lovins analysis was performed on the RSS draft report, it is remarkably relevant to the report's final version. His technical analysis reveals gross faults in the RSS's data, methods, assumptions and conclusions.

According to Lovins, the frequency of real-life reactor accidents absolutely refutess the hypothetical probabilities computed by the Rasmussen group. Lovins found that, "The RSS data and methodology yield absurd results when used to predict the likelihood of major multiple failures *which have actually occurred.*" Applying the Rasmussen techniques to one particular sequence of failures in boiling water reactors, the techniques imply that the failures would occur only once in many billions of reactor-years. "Yet," wrote Lovins, *"at least fifteen such accidents have already occurred* in the USA."

Because the Rasmussen report is probably the most important recent official reactor safety study, some readers may be interested in a more detailed analysis. Therefore, a brief overview of the Rasmussen study's shortcomings is presented below and a number of citations for further reading are included in the notes to this chapter.

Readers who find the discussion too technical can go on to Chapter 5 without any loss of continuity. However, readers who tackle the section will readily understand how the weaknesses of the Rasmussen report cast doubt upon the credibility of the AEC's entire civilian reactor safety program.

RASMUSSEN'S NUMBERS GAME

A basic reason for the unreality of the RSS results is the study's misuse of reliability and safety-analysis techniques developed by the U.S. Department of Defense and the National Aeronautics and Space Administration for the U.S. aerospace program.

One expert on those techniques is Dr. William Bryan, a mechanical engineer from the University of California, Davis. Dr. Bryan spent ten years as a reliability and safety analysis expert in the aerospace industry, where he worked on the Apollo program and on a nuclear powered rocket program. During this work, he was frequently in close consultation with the AEC on safety problems of mutual interest. Testifying before the Subcommittee on State Energy Policy of the California State Assembly, Dr. Bryan said he found:

> . . . in general, the AEC is up to ten years behind the times as far as implementing aerospace reliability and safety techniques is concerned, and as a substitute for good analysis, is pushing phony reliability and safety numbers to assure us of just the opposite. . . .

Dr. Bryan also explained that although the "fault-tree analysis" used by the RSS to compute accident probabilities could be useful in *comparing* the relative safety of mechanical systems, it could not yield meaningful probability estimates such as "one chance in a billion." Fault-tree analyses had been misused by the RSS group in their attempt to allay the public's growing fears of catastrophic nuclear accidents. The RSS used fault-trees to derive *absolute* numerical risk estimates purporting to show that the risks of serious accidents are negligible. (See box.)

The dialogue below is between Dr. Bryan and Charles Warren, Chairman of the California legislature's Subcommittee on State Energy Policy:

Chairman Warren: Can you explain fault tree analysis?

Mr. Bryan: . . . *A fault tree analysis* is where you start with some problem that can occur, some system malfunction, then you start tiering your analysis much like an organizational chart. You start with a box at the top that says you're going to have a loss-of-coolant accident. You then tier it down to the six or so things that can cause a loss-of-coolant accident, and then for each one of those six things, you analyze the things that could cause each of those six, and you just keep tiering down until you're down to the nuts and bolts of the system.

The problem in building a fault tree and getting a number out of the fault tree analysis is obvious. (You have this huge tree of possible failure mechanisms that all inter-react and all lead into other events for which you have no quantifiable data.). . . You just have to have failure rates for every point in the analysis, and there just does not exist that type of information. So you end up doing the same thing we've always done. Where you can get failure rates, you use them. . . . Obviously, a pipe used in the oil industry is going to fail much differently from one in a nuclear application, but this is the best you have got so this is what you use.

In other cases, where there is no industrial failure rate, you go back to some qualitative method or some guessing game.

If you're consistent in the use of these numbers in the fault tree, when you get done you certainly can *compare one design against another* and say this design is better than the other, if you used a common data base. . . . The absolute value of the number is totally meaningless. There is just no way that number can mean anything in terms of the real world probability of failure.

Chairman Warren: So if someone has said to me, "The likelihood of a particular event occurring is one in one thousand million," that then is really meaningless. The only time that could conceivably be meaningful is when competing systems are compared. . . .

Mr. Bryan: Exactly. . . .[5]

In addition to the methodology problems described by Dr. Bryan, the RSS results are puzzling in view of previous AEC safety research. The RSS estimates of the likeliest core melt accident, for example, are about 50 to 500 times higher than the AEC had previously stated in the 1957 Brookhaven Report and its suppressed 1964-65 update. Yet the RSS found the results of the worst accident consequences to be many times less severe than did the Brookhaven Report. This seems to imply that twenty years of AEC safety research have been guided by dubious information about which accidents we should be most wary of.[6]

To add to the confusion, AEC officials like former AEC Chairman Dixy Lee Ray have told the public that the chance of a maximum credible accident is only one in a million per reactor per year.* Even if Dr. Ray's figures are assumed correct, with 1,000 reactors operating in the U.S. as the AEC forecast for 2000 A.D., the probability of a major accident would be 3 percent, or about one in 33, over the thirty-year lifespan of those reactors. These chances cover only the estimated odds of equipment failing in a specified sequence culminating in a large escape of radioactivity. Other factors could increase the odds.

The report of Working Group 5 of the Pugwash Conference on World Affairs† advises that with 1,000 large reactors expected to be in operation throughout the world by 1990 and 3,000 by the year 2000, it would be wisest to adopt the most cautious available risk estimate — one in 10,000 per reactor-year. With 1,000 reactors in operation, the chances of a major accident then become one in ten per year. Over the lifespan of these reactors, the chances of an accident then are quite large.

In addition to its questionable results, the RSS simply did not consider some crucial aspects of reactor safety. The study did not assess the effects of an accident on the whole U.S. population, only the risks to the 15 million people within *twenty miles* of the one hundred nuclear reactors that are expected to be in operation by 1980. But radioactive pollutants do not recognize such artificial boundaries. Nor did the RSS try to evaluate the risks of nuclear

*Estimates published by experts have varied from one in ten thousand to one in a trillion.

†The Twenty-Third Annual Pugwash Conference on World Affairs met in Finland in 1973.

sabotage or terrorism. Nor, of course, did it study the major dangers inherent in producing nuclear fuel and handling nuclear wastes. In addition, the dangers of the next generation of new nuclear breeder reactors are unexplored, although the largest single item in the 1977 U.S. energy research and development budget request was the breeder. The AEC has estimated that four hundred breeders will be operating in the U.S. within twenty-five years.

Finally, Professor Rasmussen himself has acknowledged that fundamental design errors in safety systems could not be predicted by the RSS because they are currently unknown.[7] Such errors, according to Lovins, were responsible for many Atlas missile test failures and for many reactor failures.[8]

Sheldon Novick, editor of *Environment* magazine, contends that all of the RSS accident probabilities hinge upon the study's assumption about the much-disputed reliability of the ECCS. Novick writes (concerning the draft report):

> The Rasmussen report indicates that unless the emergency system functions adequately, fuel melting will result from openings as small as half an inch in the normal cooling system. Page 122 of the report gives the probability of a two-inch break in the cooling system as about one chance in 3,000 per reactor per year. The Rasmussen report does not point out what seems clear, that for 100 reactors over thirty years, the probability of such a break occurring is very high indeed. In fact, there is almost a 50 percent chance that two such breaks will occur during the lifetime of the 100 plants. Yet the report gives the probability of the fuel melting as a result of such a pipe break as one chance in 300,000 per plant per year. The difference is the assumed functioning of the emergency cooling system in 99 out of 100 accidents.[9]

But Lovins concludes that according to the RSS, "it does not make much difference in the end whether ECCS will work or not," because of the RSS new finding that small pipe breaks are much more contributors to accident risks than large ones. "If this is correct," says Lovins, "the thrust of a decade or more of official safety regulation in all countries has been wholly misdirected owing to an erroneous identification of the main source of risk. RSS and previous AEC safety analyses cannot *both* be right."

Even assuming that the ECCS works in small-scale tests or in responding to a large pipe break, the best ECCS imaginable would not prevent the destruction of a nuclear power plant in a very large earthquake. The nuclear power industry takes comfort in the infrequency of earthquakes — so much so that reactors today are being built in Class 3 earthquake zones, the zones of maximum earthquake risk. Yet no known reactor could survive the forces at the center of an extremely severe quake. In the Yakutat Bay quake of 1899, for example, the ground actually slipped 46 feet. A reactor would have been torn apart.

Reactors are only protected against quakes of specific anticipated sizes. For example, a plant might be designed to withstand a quake of 6 on the Richter scale. But if a quake of magnitude 8 occurred (one hundred times more forceful than one of magnitude 6) the plant might be seriously damaged or destroyed. The Diablo Canyon nuclear plant in California currently is being built to withstand a quake of 6.25. The 1906 San Francisco earthquake had an intensity of 8.2.

The decision on how strong a plant must be is based on predictions regarding the size of future earthquakes in the plant vicinity. These estimates in turn are based on geological studies and knowledge of the region's seismic history. But such estimates are manifestly inconclusive. Scientists have only been keeping crude records of quake magnitudes for about seventy years and earthquakes can occur in cycles of millions of years. For this reason, we do not even know how big the largest future quake may be. (See, for example *Science,* December 1975). Routine geological examination of a plant site also does not insure its safety. Many earthquake faults are not visible on the surface of the earth, yet they may cleave deep bedrock far beneath the surface on which the plant is built. Some day, it may be possible to devise an ingenious means to protect power plants and the public against every conceivable earthquake. Today reactors do not have such protection.

Much of the RSS study was done at AEC headquarters in Bethesda, Maryland. AEC specialists from headquarters as well as from the national laboratories worked on the project. Numerous subcontractors, such as Aerojet Nuclear Company and Boeing Company, also participated. It is not surprising that a study financed and conducted under the auspices of what has been the nation's greatest promoter of nuclear power should find that the risks of that technology are acceptable. No direct pressure would have been needed to achieve these results. Those involved in the study were likely to have had such strong personal and professional commitments to nuclear power that the chances of an unbiased, objective report were about the same as being struck by a meteor.

Whether a person chooses to be convinced by the Rasmussen study or not, it is undeniable that the risks of a catastrophic accident are only one small part of the total risks associated with nuclear power. And some opponents of nuclear power fear the routine hazards of processing and shipping nuclear materials even worse than they fear a terrible accident. As Chapter 5 shows, these fuel cycle risks ultimately can have *bigger* consequences than even a nuclear meltdown accident. Moreover, the chances that these neglected fuel-cycle hazards will occur are not one in a billion but 100 percent.

NOTES

1. Reactor Safety Study, As Assessment of Accident Risks in U.S. Commercial Nuclear Power Plants, WASH-1400/NUREG 75/014 Washington, D.C., 1 October, 1975.

2. "Comments by the Environmental Protection Agency on the Reactor Safety Study, An Assessment of Accident Risks in U.S. Commercial Nuclear Power Plants," Environmental Protection Agency, November, 1974; Henry W. Kendall and Sidney Moglewer (eds.), *Preliminary Review of the AEC Reactor Safety Study* (San Francisco: Sierra Club/Union of Concerned Scientists, 1974); Lovins, *op cit.;* and *Report to the APS by the Study Group on Light Water Reactor Safety, Reviews of Modern Physics,* Vol. 47, Supplement No. 1 (Summer, 1975).

3. "E.P.A. Doubts Data on Atomic Mishap," *New York Times,* December 5, 1974.

4. Amory B. Lovins, "Nuclear Power, Technical Bases for Ethical Concern" (London: Friends of the Earth Ltd. for Earth Resources Research Ltd., March 1975). Update No. 2 to 2nd edition. See note 173.

5. *Not Man Apart,* mid-May, 1974.

6. Lovins, *op. cit.*

7. Robert Gillette, "Nuclear Safety: Calculating the Odds of Disaster," *Science,* Vol. 185, September 6, 1974.

8. Lovins, *op. cit.*

9. Sheldon Novick, "Report Card on Nuclear Power," *Environment,* Vol. 16, No. 10 (December, 1974).

5

BETTING ON PERFECT TECHNOLOGY
Some "Routine" Health Effects of Nuclear Power

Dr. John W. Gofman, fifty-six, has provided courageous leader-ship and authoritative documentation for the national movement against nuclear power, launched by people concerned about its safety hazards and health effects. Grey-bearded, his gentle face crowned by a high balding dome, Dr. Gofman is a careful scientist whose piercing insights, coupled with meticulous laboratory work, have earned him a worldwide reputation. Wearing horn-rims and puffing on a pipeful of Amphora tobacco, he speaks with ease and precision. The analyses of nuclear issues he renders are cautiously framed yet often controversial and sometimes terrifying. But no one impugns his credentials.

A physician with a Ph.D. in nuclear physical chemistry, Dr. Gofman is the co-discoverer of uranium-233, and three other radio-nuclides. He has shared the Stouffer Prize, the highest American award for heart disease research, and he has published over one hundred sixty scientific papers. He was an associate director of the Lawrence Radiation Laboratory from 1963-1969 and is now an Emeritus Professor of Medical Physics at the University of California, Berkeley.

Interviewed in 1975, Dr. Gofman recalled the early days of the nuclear weapons program when officials repeatedly assured us that it would have no public health consequences. "You don't have to worry about the radioactivity produced," the officials said, "it'll all go up in the stratosphere and never come back."

"Then," observed Dr. Gofman, "worldwide fallout was discovered, and that came as a *surprise*, you know, they didn't expect that. And in fact, when they started to find out [it] was coming back, they made a highly secret project in the early fifties called 'Project Sunshine,' to measure around the country and the world, how much strontium-90 and cesium was getting back."

"Then, we always had reassurances that, okay, so it did fall out, but there's a safe threshold [of exposure to radioactivity] so that you won't have any health effects. And we had in 1954, largely in support of the weapons program, the statement of the National Committee on Radiation Protection that a tenth of a rad per day wouldn't have any physical effects — that 36 rads per year would be safe. Now we're arguing about .17 rads per year . . . the numbers that were stated to be safe had to be kept moving down as more data developed."

As summarized by Dr. Gofman, the public pronouncements sound as though AEC officials were suffering from a classic case of foot-in-mouth disease: "The radioactivity can't come back. Oh, but oops, it is coming back. But the amounts are small and you can tolerate so much, because there's a threshold." In no small measure due to a celebrated controversy between the AEC and Dr. Gofman, who was joined by his colleague Dr. Arthur R. Tamplin, we now know that the existence of such a threshold is a pipedream.

And as a result of recently completed "first approximations" by Dr. Gofman on the health effects of radioactive fallout and reactor plutonium, the enormous consequences of the AEC's "mistake" are only now becoming apparent. The Gofman results are a scientific bombshell whose shockwaves have barely begun to reach the scientific community. They imply that more than a hundred thousand people have been committed to die from lethal lung cancers due to the weapons tests, and that more than twenty-five million people will meet the same fate from reactor plutonium as a result of nuclear power expansion.

PLUTONIUM AND CANCER

Working unassisted in his medieval-style San Francisco townhouse and far afield from his once-promising cancer/chromosome research, Dr. Gofman completed a study in May 1975 on the cancer hazard from inhaled plutonium.[1] Two months later, using baseline calculations from that study, he estimated the lung cancer effects produced by nuclear weapons tests.[2]

Despite what Dr. Gofman calls "critical voids in mankind's knowledge" about plutonium dosimetry,* he was able to calculate that lung cancers from exposure to plutonium were 127 times as likely in smokers as in nonsmokers. (The fatality rate among lung cancer victims is about 95 percent.)† Using a new and improved model of plutonium behavior in the lung along with the fact that reactor plutonium is 5.4 times more hazardous than plutonium in weapons fall-out,[3] Dr. Gofman concluded that .011 micrograms‡ of reactor plutonium deposited in the lung is enough to produce lung cancer in a cigarette-smoking male. That, by the way, is 385 trillionths of an ounce. One pound of reactor plutonium, therefore, contains more than 42 billion of such lung cancer doses!#

This conclusion plus other data about plutonium toxicity imply that if the general U.S. population is exposed to the current federal limits on plutonium exposure, more than seven million *extra* lung cancers (above the spontaneous cancer level) can be expected over a thirty-year period among males in that generation.§ On an annual basis, this amounts to 235,000 extra lung cancers — roughly a 400 percent increase.

Dr. Gofman writes, "Many serious public health experts con-

---

*The process or method of measuring radiation dose.

† Several prestigious but official or semiofficial scientific research groups including the British Medical Research Council (BMRC),[4] the National Academy of Sciences Committee on the Biological Effects of Ionizing Radiation (BEIR),[5] and the International Council on Radiation Protection (ICRP)[6] somehow had missed this essential fact in their studies on plutonium. Both the BMRS and the BEIR committee had used patently erroneous and grossly oversimplified models of plutonium behavior in human lungs. Both failed to acknowledge the bronchi as the true site of most lung cancers and overlooked the differential clearance rate of plutonium in the lungs of smokers compared to nonsmokers.[7]

‡A microgram is a millionth of a gram.

Correspondingly, for nonsmokers, the minimum lung cancer dose is 1.4 micrograms of deposited plutonium, and each pound of reactor plutonium contains 338,000,000 of these doses.

§The spontaneous lung cancer rate among women is about a quarter of the spontaneous rate in men, but smoking habits differ, so several adjustments to the Gofman data would be necessary to use his model in forecasting the incidence of additional lung cancers among women.

sider [our] 63,500 lung cancer fatalities per year to represent a most serious epidemic. How should they view the burgeoning plutonium-based nuclear fission energy economy, proceeding under regulatory standards that would permit a four-fold increase supplementary to this epidemic?"

For workers in the country's rapidly expanding plutonium industry, Dr. Gofman's findings are likely to be a shock: the currently allowable exposure guidelines for plutonium workers permit each cigarette-smoking worker to accumulate during his or her lifetime, four and a half times the amount of reactor plutonium capable of causing cancer. For the nonsmoking male worker, exposure to the allowed levels generates a one in thirty chance of lung cancer per lifetime.*

The AEC was fond of reassuring the public that little of the allowed radiation doses would actually be delivered, yet they resisted lowering those allowed limits. Based on AEC forecasts of nuclear power growth,† a full-scale nuclear industry would produce 440 million pounds of plutonium during the next fifty years. Clearly, even the release of a small fraction of that amount would greatly exceed legal limits. And in that event, unsecured AEC and industry promises about low release levels would be stone cold comfort.

Dr. Gofman has calculated that even if the nuclear industry contains its plutonium wastes 99.99 percent perfectly — losing only one part for every ten thousand handled — then an average of five hundred thousand extra lung cancer deaths per year can eventually be expected.‡ That is *a necessary consequence* of expanding the nuclear fission program as currently planned. Over fifty years, by these calculations, more than twenty-five million people would die. Of course, there are many other toxic products in reactor wastes, and these too, would add to the human death toll.

---

*These results, Gofman shows, are quite consistent with existing studies in which cancers have been induced by radiation in laboratory animals.[8]

†The AEC in *Nuclear Power 1973-2000* (WASH-1139 [72]) forecast that U.S. nuclear electrical generating capacity would be about one thousand two hundred billion watts, which would be the output of a thousand or more large nuclear plants. Many analysts, observing recent setbacks to the nuclear industry, now regard this as unlikely. Yet even with five hundred plants in operation, the nuclear industry would still produce hundreds of thousands of pounds of plutonium a year.

‡None of the Gofman computations depend on the controversial "hot particle theory" of Arthur R. Tamplin and Thomas B. Cochran.[9]

Physicists Thomas B. Cochran and Arthur R. Tamplin of the Natural Resources Defense Council, Inc. have asserted that the release of intensely radioactive alpha-emitters, such as plutonium, which they termed "hot particles," from the nuclear power and nuclear weapons industries, requires that federal radiation exposure standards be greatly tightened to protect the public from high cancer risk. As regards plutonium, the hot particle theory implies that a single speck of plutonium oxide delivering a dose of only .3 millirems when its radiation is averaged over the entire tissue mass of a human's lung, could deliver *4,000 rems* to the tiny portion of the lung tissue actually irradiated by the plutonium. The results of Drs. Cochran and Tamplin (and the independently derived results of Dr. Gofman) all clearly imply that an individual receiving far, far less than the allowable NRC whole-body exposure guidelines (170 millirems) in the form of a plutonium oxide particle in the lung might easily be receiving a fatal dose of radiation. For this reason alone, the nuclear power industry's reassurances that they will not deliver the full 170 millirem dose to each member of the general public is no assurance whatsoever from a safety standpoint.

As a result of Cochran and Tamplin's research, NRDC in 1974 petitioned the AEC and the Environmental Protection Agency to make federal radiation exposure standards 115,000 times more strict. These proposed rule changes have not been adopted.

In his analysis of lung cancers resulting from the weapons fallout which the AEC had claimed "would all go up in the stratosphere," Dr. Gofman found that approximately 900 pounds of plutonium had been deposited in the U. S. through 1972 as a result of weapons testing in the 1950s and 1960s. He then estimated the amount of this plutonium which had found its way into human lungs and calculated what its effects would be, based on his knowledge of plutonium toxicity.

Dr. Gofman found that, to date, roughly 116,000 persons have already been committed to lung cancers in the U.S. and about a million have been doomed throughout the northern hemisphere. Roughly one thousand of these cases a year are already occurring in the U.S., he concluded, and ten thousand are appearing throughout the hemisphere. Dr. Gofman commented, "But since the lung cancer cases caused by plutonium exposure do not carry any flag

that tells us that these particular cases are the ones caused by plutonium exposure, the absurd statement is possible that 'I don't
know anybody that's died as a result of exposure to plutonium, do
you?'"[10]

Had medical results of such importance been derived by anyone of less stature than Dr. Gofman, they would probably have
been greeted with disdain by nuclear advocates. But instead, Dr.
Gofman's work has been respectfully cited by major U.S. newspapers such as the *New York Times*, the *Wall Street Journal*, and
others. He also has substantial support in the scientific community
and his work has been hailed as a major contribution to human
welfare by distinguished Harvard biologist, Professor John T. Edsall.
Dr. Gofman's work also merits special attention because, several
times in the past, when many other "experts" were serenely accepting AEC safety platitudes, Dr. Gofman was already denouncing
those claims — at the risk of his position and reputation.

For example, Dr. Gofman blew the whistle when the AEC
tried to sell the public the idea of a "safe threshold" of radiation
exposure.[11] He was on leave from most of his teaching duties at
the University of California in 1963-1969 and was Associate Director of the AEC-supported Lawrence Radiation Laboratory at Livermore, California, heading its newly formed Biomedical Division.
Along with Dr. Tamplin (an AEC employee at that time) and
others, Dr. Gofman was trying to find out the relationship between
cancer and radiation. Their results led them to predict that every
unit of radiation would cause twenty times as much cancer as
expert bodies such as the International Commission on Radiological Protection had said three years before.

"Almost immediately after saying it," Dr. Gofman recalls,
"we had a deluge of vitriolic comment from the nuclear reactor
vendors and from the electric utility industry — and from the AEC,
which struck us as somewhat unexpected. . . . After all, we thought
it was good for them to know about it."

Finally, Dr. Gofman realized that the nuclear advocates were
on the course of selling the idea that there was a "threshold" or
safe level of radiation exposure below which the nuclear industry
could operate without harming anyone.

But Dr. Gofman and Dr. Tamplin had surveyed the literature
on low-dose radiation, including AEC studies, and conducted re-

search of their own. A clear and unequivocal linear correlation emerged between cancer and radiation, extending even to tiny doses. (This finding is now widely accepted by scientific organizations.) The researchers also concluded that if the U.S. public were actually exposed to the allowable radiation limits set by the AEC, 32,000 additional cancer, plus leukemia deaths, and a large number of additional genetically-induced deaths, would result.

The AEC and others convened a National Academy of Sciences committee to study the problem. The committee's study, known as the BEIR report, found the two researchers' figures might be four or five times too high, but definitively agreed there was no safe exposure threshold. Despite this dispute over the number of cancer cases, the BEIR report did affirm the principles on which the Gofman-Tamplin work was based.

Dr. Gofman and Dr. Tamplin also tangled with the AEC on another aspect of the same issue. In 1969, they were asked by the commission to scrutinize some alarming data just published by Dr. Ernest Sternglass, a University of Pittsburgh physics professor. Dr. Sternglass had attributed four hundred thousand infant and pre-natal deaths in the U.S. to atmospheric nuclear tests in Nevada in the fifties. (A few years later, he was to claim a correlation between emissions from certain Pennsylvania nuclear power plants and increased infant mortality nearby.)

To the AEC's dismay, instead of totally refuting Dr. Sternglass' fallout research, Dr. Tamplin concluded that although Dr. Sternglass' estimates were perhaps one hundred times too high, about four thousand human infants had actually died from the Nevada fallout.

Although they were working under AEC grants, the two researchers clung tenaciously to their views, despite strong AEC pressure to gracefully bury that result and follow official policy. Dr. Gofman, for example, presented their results at a meeting of the Institute for Electrical and Electronic Engineers; in low-keyed terms, he recommended lowering allowable radiation exposure limits by a factor of ten.

Soon, Dr. Tamplin's research funds were reduced from $300,000 to $70,000, forcing him to cut his staff from thirteen to one. In attacking Dr. Gofman, the AEC was more circumspect. They did not directly interfere with his participation in congressional committee hearings or state-level hearings on nuclear power. Nor did they overtly intercede against the articles he wrote about nuclear power during 1970-1972. Nor did they fire him, despite his public statements. But they did find a way to repay Dr. Gofman for his heresy.

In 1972, about three years after Dr. Gofman's outspoken opposition to nuclear power began, the AEC recommended to his supervisor that funding for Gofman's work end. Dr. Gofman had been conducting a very active and promising quarter-million-dollar program of research on the relationship of chromosomes and radiation with cancer.

He then turned to the National Cancer Institute whose director assured Dr. Gofman that the Institute thought highly of his cancer-chromosome work. But after a puzzling two-month delay, the Institute spurned his request for funding.

Once his resignation from Livermore was effected, and his research program was destroyed, the AEC did not remove that $250,000 from the Laboratory's budget.

It is ironic that had the AEC not acted vindictively against him, Dr. Gofman might today be conducting chromosome research in a well-equipped laboratory and teaching at the University of California, from which he has now taken an early retirement. He might then never have written his recent seminal papers on the health effects of plutonium.

BELATEDLY DISCOVERED HAZARDS

For years, controversy over nuclear power has often focused on the hazards of reactors at plant sites or on relatively peripheral issues such as thermal pollution.* This myopia suits nuclear advocates just fine; they are fond of looking at one small phase of the nuclear fuel cycle and saying, "Under normal operation, this won't produce much radioactivity, so we have no problem with nuclear power."

But as mentioned earlier, the nuclear fuel cycle includes uranium mining, uranium conversion, uranium enrichment, fuel fabrication, radioactive scrap recovery, fissioning uranium in reactors, waste reprocessing, waste storage, and many transportation links. Each of these stages bears risks of varying degrees. Until now, nuclear advocates have conveniently overlooked or dismissed the hazards in the early stage of the fuel cycle, before the uranium is fissioned. Even many sophisticated nuclear opponents have been misled by this ploy. Consequently, virtually the only early-stage fuel cycle hazards to receive public attention are the lung cancer epidemic among uranium miners[13] and the radiation from radon gas in uranium mill tailings.[14] But recently, new evidence has appeared that dwarfs all previous estimates of public health risks at the front of the fuel cycle.

Calculations by physicist Robert Pohl of Cornell University now indicate that radiation from the element thorium-230 in uranium tailings can produce *many millions* of deaths if the nuclear power program continues unabated.[15] Moreover, adequate measures to control against thorium release are probably unfeasible in the long run for physical as well as economic reasons.

Before Pohl's work, the primary concern about uranium tailings had been their radium, a radioactive element found naturally with uranium ore. As radium ages, it releases radon, a chemically inert gas which, like radium (a solid), is a lung cancer threat. In a

---

*Beneath the heading, "Radioactivity from Nuclear Plants is Low," pro-nuclear literature prepared for free distribution by the Atomic Industrial Forum attempts to focus attention primarily on this environmental problem by stating, "Light water reactors show up least favorably in one major area. They are thermally less efficient and therefore a greater flow of cooling water than fossil fuels (sic)."[12] The Atomic Industrial Forum is the principal lobbying group for the nuclear industry.

natural ore body underground, layers of earth and rock greatly slow the release of radiation and radon gas into the environment. But when the ore is milled into finely crushed tailings sand so that uranium and thorium can be extracted, both radium and radon can escape from the tailing heaps with great ease.

When present in finely ground tailings, radium easily dissolves in water. Although major uranium mining activity began in the southwestern United States in 1946, stream and river pollution by radium was not officially noticed until the late 1950s and the dumping of radioactive liquid mill effluents in Colorado was not stopped until 1959. By then, contamination severe enough to kill river fauna poured into water systems from which people drank and irrigated crops near Aztec and Farmington, New Mexico.[16] Radiation levels were well in excess even of federal guidelines. The mine tailings were also used in the construction of thousands of homes in the Southwest, both as landfill, and as a substitute for sand in concrete.

It was later discovered that radon in the tailings could penetrate concrete, masonry, and mortar. Thus, occupants of structures built with tailings could receive potentially lethal doses of radiation — without ever seeing the cause. Although the AEC was aware of a possible danger from the use of tailings in construction as early as 1961, the AEC disclaimed responsibility for controlling the tailings after they left the mill and apparently did not adequately alert state officials to the consequences of neglecting the problem. Not until 1966 did Colorado Public Health officials become concerned about unusually high levels of radiation in certain buildings. Yet it took the AEC another three years even to restrict public access to tailing piles.

More recently, a 1975 EPA survey of water supplies in the Grants uranium mining and milling area of New Mexico found much of the area's water to be infiltrated with water from uranium tailing ponds and with contaminated mine drainage containing radium and uranium.* Among the company's accused by the EPA of pollution in the area were Anaconda Copper Corp., United Nuclear Corp., and Kerr-McGee Corp.†

---

*U.S. Environmental Protection Agency, *Water Quality Impacts of Uranium Mining and Milling Activities in the Grants Mineral Belt, New Mexico,* EPA, Region VI, Dallas, Texas 75201, September 1975.

†For a valuable first-hand series of articles on uranium mining's impact on the

The tailings hazard examined by Professor Pohl is of a far more serious nature than any tailings problem known so far. Roughly 100 million tons of uranium tailings have been deposited in large and generally unstable piles on the surface of the earth in the Southwest. These enormous mounds, sometimes blanketing hundreds of acres, are poorly protected from the erosive forces of wind and water. Some are uncovered; those that are covered lie beneath no more than two feet of ordinary soil.

And about twenty billion tons of tailings — enough to cover Rhode Island completely to a depth of seven inches — will be produced by the year 2000 if the nuclear industry grows as planned. [17] The piles contain large quantities of radioactive thorium-230, which gradually disintegrates to produce radon-222, polonium-218, polonium-214, lead-214 and other elements.

Because the half-life of thorium is 80,000 years, it will remain dangerously radioactive for 800,000 years.* Thus even though only small quantities leach into the environment at once, the cumulative effect of its steady seepage will be immense.

Tallying the volume of uranium tailings that will be produced if the U.S. nuclear industry expands as planned, Pohl's work implies that over a thirty-year period, an average of more than a thousand people per year would die from thorium-induced cancers. But the annual effects would keep accumulating until more than 11 million people had perished. These casualties would not be limited to the U.S. Should all these people die so we can have nuclear power today?

How is it that thorium-230, a public health menace of colossal

---

people of rural New Mexico, particularly the Pueblo Indians of Paguate, N.M., see Tom Barry's feature series in the Albuquerque, New Mexico alternative newspaper, *SEERS*: "Uranium Boom in Grants," (December 13-20, 1975); "Gulf Oil Digs Into Mt. Taylor," (December 20-January 16, 1976); "The Future of the Jackpile, Voices of Paguate," January 31-February 14, 1976.)

*Radioactive elements decay at characteristic and unvarying rates. Their half life is the time required for each such element to lose half of its initial radioactivity. If there are x atoms of a radioactive element at any particular instant, there will be $\frac{x}{2}$ atoms present after one half life has passed, $\frac{x}{4}$ after two half lives, etc. The passage of ten half-lives are therefore required to reduce the radioactivity of a substance to a thousandth of its original level. Ten half lives of thorium equal 800,000 years. As a rule of thumb, the passage of 20 half lives is considered necessary before a radioactive substance can be considered to have become "safe."

scale, simply went neglected, unnoticed, or untold? How many other grotesque surprises will be discovered as the full consequences of the radioactive fuel cycle become known? And how much faith can be put in guardians of our public health and safety who have made little mention of this enormous threat?

The thorium story is not an isolated instance in the nuclear power program. The federal radiation regulatory bodies have time and again failed to protect the public from the abuses of the nuclear industry. Characteristically, the regulators have not recognized crucial threats to public health until after the threats have been created.

For example, the Environmental Protection Agency has recently proposed to lower the allowable exposure levels from nuclear fuel cycle radiation to a twentieth of current levels.*[18] Yet in that proposal, the EPA projects that previously neglected emissions of the radioactive pollutants tritium and carbon-14 will produce more *than sixty times the health effects*† over the hundred years following their release than the total of all other long-lived and short-lived radioactive pollutants studied by the agency! And although the EPA proposed limits on the releases of radioactive krypton-85, iodine-129, plutonium-239, strontium-90, and cesium-137, it proposed no *specific* standards for tritium or carbon-14, for which pollution control technology does not exist.

Apparently, the EPA did not even discover the carbon-14 hazard until 1974. And the agency did not project the health effects of carbon-14 beyond the next hundred years after its release, although carbon-14 has a half-life of 5,730 years and, as we have seen with thorium-230, the first hundred years may be only the "tip of the

---

*Currently, the federal radiation protection standards limit exposure for the general public from the nuclear fuel cycle to 500 millirems for the whole body and 1500 millirems for the thyroid. More stringent exposure *guidelines* which do not have the force of law have been recommended by the Federal Radiation Council. These set public exposure limits of 170 millirems for whole body exposure. A millirem is equal to one thousandth of a rem. (See Chapter 2, Note 3 for more details.)

†The EPA defined "health effects" as limited to total cancers including leukemias, and serious genetic diseases. They specifically and cavalierly excluded other possible radiation effects, such as "the genetically-related component of diseases such as heart disease, ulcers, and cancer as well as more general increases in the level of ill health . . . " along with radiation effects on growth, development, and life spans.[19]

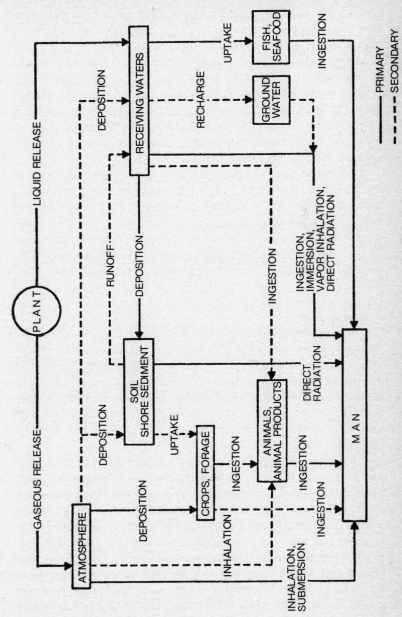

Figure 5-1. Pathways for External and Internal Exposure of Man

iceberg." In addition, these EPA projections assume *normal operation* everywhere in the fuel cycle. Yet the entire history of the nuclear power program is filled with *thousands* of abnormal occurrences,[20] and many excessive radiation releases. One such uncontrolled release can be vastly larger than all routine releases combined. For example, a plutonium leak at the AEC's nuclear weapons factory at Rocky Flats, Colorado, dispersed much more plutonium "than the integrated effluent loss during the seventeen years of plant operation. Tens to hundreds of grams of plutonium went offsite, ten miles upwind from Denver."[21]

Given this history of flagrantly inadequate regulatory action by the AEC and even the EPA, it is not reassuring to see the nuclear industry threatening to embark on two new and ultra-hazardous fuel cycle undertakings. They are the reprocessing of "spent" (fissioned) nuclear fuel, and the commercial use of recovered plutonium as fuel for reactors. Both these activities would necessitate the eventual handling of millions of pounds of plutonium and other high-level wastes. Moreover, these steps will be necessary to fuel the now-experimental breeder reactor, if it is ever to enter into commercial use.

Why are breeder reactors and fuel reprocessing desired by the nuclear industry, and what additional new hazards will they create? When uranium fuel rods are fissioned in an ordinary reactor, the amount of fissionable uranium-235 diminishes as other radioactive elements accumulate in solid, liquid, and gaseous form. Consequently, about a third of the one hundred tons of uranium fuel must be removed from a reactor and replaced every year in a refueling operation.

A commercial reprocessing plant will take in rail or truck shipments of spent fuel rods in huge fuel casks. Machines will open them in a shielded chamber where the rods will be chopped into short sections and dissolved in vats of acid. By chemical processes, unused uranium-235 and plutonium-239 will be extracted from the resulting hot acid "soup." The uranium and plutonium will then be available for use in fabricating mixed uranium and plutonium reactor fuel, which today can only be used in test reactors. Both the irradiated fuel elements and the waste soup they yield are known as high-level waste.* The soup contains highly radioactive

---

*Low-level wastes are equipment, tools, clothing, or anything that has become

fission products like strontium-90 and cesium-137, which remain poisonous for hundreds of years.

The reprocessing plant must not only successfully recover re-usable radioactive material, but must also solidify its high-level waste into a ceramic or glass-like substance for shipment to a waste storage facility. All this must be done without exceeding the safe limits set for discharges of radioactive liquids, solids, and gases into the environment. Millions of curies of radioactivity must be handled and almost perfect containment must be maintained, if severe public health consequences are to be avoided.

Currently, there is no commercial fuel reprocessing occurring in the U. S., and only one such plant, Nuclear Fuel Services (NFS) in West Valley, New York, has operated. Before its closure in 1971, the plant emitted excessive amounts of radioactive gases and liquids, contaminating surrounding air, water, fish, and wildlife, including deer.[22] The NFS facility plant is now being renovated and expanded to handle fifteen times as much nuclear waste as before, but its previous record gives no cause for confidence in its future safe operation. However, despite renovations, the plant may never reopen. The Natural Resources Defense Council has reported that Getty Oil is trying to sell the plant but has yet to find a buyer.[23]

One other reprocessing venture, the Barnwell facility near Aiken, South Carolina, is officially scheduled to begin operation in 1978. For reasons that are easily understood, a five-year battle has been going on between Allied-General Nuclear Services, builder of the plant, and environmentally concerned opponents. The plant, after five years of operation, will contain as much radioactivity in storage as would be produced in a full-scale nuclear war; its liquid waste tanks will hold many times the fission product radioactivity produced by all atmospheric nuclear weapons testing combined. Allied-General officials are expecting eventually to handle five hundred

radioactive during the fuel cycle, not including high-level wastes. About four million cubic feet of such contaminated materials, encased in wooden containers or steel drums, already have been buried in shallow trenches at just one government-run site in Idaho. Plutonium-bearing solids may also be included in low-level wastes. The "level" of a waste is an indication of the penetrating power of its radiation, not of its toxicity.

shipments of radioactive fuel each year — casks of such phenomenal toxicity that "they will give off enough radiation in transit to
give anyone near one of them for only seventeen hours his allowable
radiation dose for an entire year."[24]

According to journalist Amanda Spake, "*The company itself*
predicts in its environmental reports that railroad accidents like
this fall's West Virginia derailment and fire, could release fission
wastes that would contaminate food crops and land along the train
routes. *'It is conceivable*,' Allied-General reports note, *'that such
an accident might occur every eight years*.'"[25] Residents near Aiken
S.C., have plenty to worry about in addition to shipping cask accidents. The Barnwell plant will routinely be releasing iodine-129,
for instance, a substance with a half-life of 16 million years, that
lodges in human thyroid glands.

Skepticism about Allied-General officials' ability to hold down
toxic releases to officially prescribed levels seems well justified,
considering the EPA's proposed plutonium emission standards for
reprocessing facilities. By these standards, Barnwell together with
any future reprocessing plants would be allowed to release no more
than *eight thousanths of a gram of plutonium per year* while reprocessing fuel from the whole nuclear industry. Because the Barnwell plant will be handling hundreds of tons of plutonium by the
1980s, such low emissions would require the plant to achieve a
containment standard of one part in a billion! Comments Dr. Gofman, "When somebody tells me, 'We're going to do something
year in, year out, day in, day out, to a perfection of one part in a
billion,' I think they're simply insane."

The record of the early efforts at plutonium fuel fabrication
shows just how poor containment has been in this dangerous fuel
cycle activity that is nonetheless much safer than reprocessing. In a
fuel fabrication plant, reprocessed plutonium is formed into pellets
and inserted into reactor fuel rods. Numerous instances of plutonium contamination of workers and facilities at the nation's plutonium fuel factories have been reported:

> The record reveals a dismal repetition of leaks in glove boxes; of
> inoperative radiation monitors; of employees who failed to follow
> instructions; of managers accused by the AEC of ineptness and
> failing to provide safety supervision or training to employees; of
> numerous violations of federal regulations and license requirements;

of plutonium spills tracked through corridors, and, in half a dozen cases, beyond plant boundaries to automobiles, homes, at least one restaurant, and in one instance to a county sheriff's office in New York.[26]

Four companies have contributed to this accident roster: Nuclear Fuels Service, Inc., Nuclear Materials and Engineering Corporation (NUMEC), Gulf United Nuclear Fuels, and Kerr-McGee Corporation. Of these four, two are no longer fabricating plutonium fuel, and the others have had serious safety problems. Gulf United Nuclear Fuels of Long Island abandoned plutonium fuel fabrication "after a fire and explosion on 21 December 1972 injured one worker, contaminated two, and according to the AEC's investigative report of the accident, 'grossly contaminated' a working area with plutonium."[27] Nuclear Fuels Service, Inc., a Getty Oil subsidiary, closed its plant the same year for expansion and decontamination; had the plant not shut down, the AEC might have forced it to do so because of its numerous safety violations.[28] The two fuel fabrication facilities currently operating are Nuclear Materials and Engineering Corporation (NUMEC) near Pittsburgh and Kerr-McGee at Cimarron, Oklahoma, where seventy-three workers received internal plutonium contamination over a four-year period. Violations at the Kerr-McGee plant attracted national attention,[29-34] especially following the tragic case of a young Kerr-McGee laboratory technician named Karen Silkwood.

Silkwood, age twenty-eight, became deeply concerned about safety conditions at the Kerr-McGee Cimarron plant after inhaling plutonium there in July 1974 as a result of the company's failure to provide her with the proper size respirator. Silkwood subsequently became active in the Oil, Chemical, and Atomic Workers' Union and complained to union officials about plant safety hazards and about alleged falsification of fuel rod quality control tests.

At the suggestion of union officials, Silkwood began covertly compiling a dossier in a manila folder to document the alleged test falsifications and doctoring of fuel rod x-rays. She spent weeks staying late after work building her case. These activities were observed by plant personnel and, not long after, in November 1974, Silkwood discovered that she somehow had become internally contaminated by plutonium. Because there had been no recent accident at the Kerr-McGee plant, Silkwood had her apartment checked by company radiation investigators. In her refrigerator, they found a package of cheese and a package of bologna, *both contaminated with plutonium.*

Panic-stricken but unwilling to give up her battle against the company, Silkwood with her manila folder of evidence was on her way to meet with *New York Times* reporter David Burnham and an OACW union official on November 13, 1974 when her car left the road and crashed in a culvert. Silkwood was fatally injured in the crash. The Oklahoma State Patrol listed her death as accidental but an auto-crash expert hired by the OCAW union concluded that the car had been rammed off the road from behind by another vehicle. Silkwood's manila folder was not found among her personal effects in the car.

A subsequent AEC investigation of Silkwood's charges resulted in findings that Kerr-McGee endangered the health of workers, manufactured faulty fuel rods, and falsified inspection records. Another investigation into Silkwood's death and into health problems in the plutonium industry has been scheduled by a Senate Government Operations subcommittee chaired by Senator Lee Metcalf (D-Montana). Metcalf will attempt to resolve doubts expressed by the National Organization of Women about the adequacy of previous investigations by the Nuclear Regulatory Commission, the FBI, and the General Accounting Office.

The contamination of workers is significant not only as an industrial safety issue but also as a public health danger in that irradiated workers continue to contribute to the genetic pool and can thereby pass on radiation-damaged genes to future generations. Workers are allowed ten times the whole body radiation that a member of the general public can legally receive. The International Commission on Radiation Protection has reported that, from the standpoint of the genetic damage produced by radiation, it makes no difference in the increased rate of mutations in a population whether radiation-damaged genes are introduced into the population by a few individuals with a high rate of damage or by many individuals who have received small doses. Dr. Gofman has calculated that with a thousand 1000 MW plants assumed operating, their routine radiation will give nuclear plant workers a radiation dose that is the *genetic equivalent* of giving 8 millirems to the entire population. That is eight times the allowable dose limit recommended by the BEIR study and it is sixteen times the dose the nuclear power industry has said the public will get from a fully developed nuclear economy.

In addition to advocating the legalized reprocessing of nuclear fuel, the nuclear industry also wants governmental approval for the commercial use of recovered plutonium as reactor fuel. The Nuclear Regulatory Commission has deferred final action on this proposal until 1977 or 1978, but apparently has encountered extreme pressure from the nuclear industry to approve it. At this writing, the NRC has agreed to issue "interim licenses" for plutonium recycling and is currently being sued by the Natural Resource Defense Council and other environmental groups to prevent recycling.

The nuclear power industry is concerned that its rapid expansion will, within a few years, exhaust the U. S. supplies of high-grade, low-cost uranium ores. Then the industry would have to resort to low-grade Tennessee shales. Because these shales have extremely low uranium contents, great masses of rock would have to be processed at a much higher economic cost and with an even greater production of uranium tailings than for current uranium mining operations.

Already the price of uranium has begun to zoom. Uranium which a few years ago was selling for only $6 a pound is likely to cost at least $30 in 1978. If the cost of uranium were not such a small fraction of the total costs of producing nuclear power, the tremendous price hikes by now might have brought the industry to a halt. Nuclear power's biggest economic advantage has been the low cost of its fuel, and some observers of the industry believe that without the use of plutonium as a fuel, the industry could not ultimately survive.[35]

But according to a study by the Natural Resources Defense Council, plutonium recycling could only reduce uranium requirements by a marginal 10-15 percent, reducing the overall cost of nuclear power by 2-3 percent.[36] Therefore the nuclear industries now lobbying for plutonium recycling want it as a means of generating fuel for new "liquid metal fast breeder" reactors, the most hazardous kind of reactor ever developed.

Unlike the current reactors, breeders are fueled by plutonium as well as uranium, and are designed to produce more plutonium than they use. The breeder does this by bombarding the common uranium isotope, uranium-238, transforming it to plutonium-239,* or by converting thorium to uranium-233. The later is a toxic element with a 160,000 year half-life.

Even without the breeder, plutonium recycling adds gigantic hazards to the fuel cycle. Whereas the conventional uranium-235 fuel used in a reactor is not of weapons grade, plutonium fuel is — and less than five pounds of plutonium is enough to make an atomic bomb.[37] If the plutonium produced in ordinary uranium-fueled reactors is not recovered in a reprocessing plant, it remains in spent fuel rods which are heavily poisoned by toxic products and are difficult to remove from a reactor without massive shielding. Therefore, the weapons materials are relatively protected from diversion for use in illicit bombs.† But the plutonium fuel rods for breeder reactors can be diverted much more easily. Fresh unused plutonium fuel rods contain weapons-grade plutonium that has not yet been heavily contaminated by the reactor's fission products. Consequently, these rods are much easier to steal, and their contents are much easier to process.

In virtually every respect, breeder reactors are more complex, costly, and dangerous than present reactors, whose environmental hazards they greatly intensify. They will require more concentrated nuclear fuel than conventional reactors, and are expected to hold one to four tons of plutonium in their cores, enough to kill everyone on earth. Because their fuel is not only more radioactive, but also is held in a smaller area, any overheated fuel, if it melts, can reassemble into a supercritical mass. This would result in a nuclear explosion. The utilities who reassure their customers that nuclear reactors cannot explode like a bomb evidently were not thinking of breeder reactors.

To keep their fuel from melting without slowing down the fast atomic reactions necessary to breed nuclear fuel, the kind of breeders now under most intensive development are cooled by hot molten sodium instead of by water. But use of sodium as a

---

*Uranium-238 is more than 100 times as plentiful as the uranium-235 now fissioned in ordinary reactors. By using uranium-238 instead of uranium-235, breeder reactors would draw on a nearly unlimited fuel supply, thereby cutting nuclear fuel costs.

† The problem of safeguarding plutonium also implies serious threats to civil liberties. These two hazards are discussed in Chapter 9, Nuclear Insecurity.

coolant introduces serious new hazards because sodium will cause a powerful explosion if it contacts air or water, and it reacts vigorously with metal structures to weaken them.

Even if a breeder reactor is properly cooled initially, fuel rod failures could cause a fatal reactor malfunction.

The rods can be damaged by the intense radiation in the breeder's core that can cause them to swell, become brittle, and eventually break.*This could lead to some blockage of coolant flow. Fuel rods with manufacturing defects are particularly susceptible to failure. Those made at the Kerr-McGee plant at which Silkwood worked were intended for use in a new experimental breeder reactor being built at Hanford, Washington, under an AEC contract. Commenting on the fuel rod danger, Professor Henry W. Kendall of the Massachusetts Institute of Technology has asserted that, "These failures could start off an accident which would result in the release of huge amounts of radioactivity."[38]

Due to the extreme hazards of breeders, and growing doubts about whether they will ever be economical, breeders may never become commercially available in spite of AEC's projection that four hundred of them will be in operation in the U.S. by 2000.

Yet despite the uncertainty of breeder viability, the nuclear fission industry was launched on the premise that breeders *would* work — and produce useful power by the 1980s." Now — even officially — they are not expected to be commercially useful until the 1990s. According to *Science* editor Allan V. Hammond, et al, "LWRs [the kind of reactor in use today] were never considered as more than *a stopgap* by the early prophets of nuclear power; commercial utilization grew out of the successful effort to develop nuclear power plants for submarines. If fission is to become a major source of energy, breeder reactors will be needed. . . ."[39] Such are the irreconcilable contradictions on which the nuclear fission program is based.

The breeder's limited operating history is absolutely no cause for optimism about its future safety or economy. Two breeders have already been built to produce electric power. The Fermi reactor eighteen miles from Detroit was in a preliminary phase of operation when a serious meltdown occurred. The reactor was relatively

---

*This effect was discovered only a few years ago.

small and was only running at a tenth of its power so most of the radioactivity released was held within the containment shell. Experts later agreed that a catastrophic reassembly of the fuel into a critical mass leading to an explosion had been narrowly averted. The Fermi reactor was later permanently "decommissioned." A Soviet breeder reactor built on the Caspian Sea, which began operation in the summer of 1973, had to be shut down in less than six months. "Water reached the liquid sodium primary coolant, there was an explosive termination of coolant flow, a meltdown ensued, and the plant will not be in operation before three to five years. . . . "[40]

Despite the breeder's shabby performance record, it has been the AEC's and ERDA's top energy research priority in recent years, sapping a total of more than two billion dollars already.[41] Current spending is in the neighborhood of $500 million annually, and the two major breeder projects now under way are rife with cost overruns.* Additional spending of more than $6 billion has been planned. Yet the economic benefits of this environmental menace remain to be demonstrated.[42,43]

In addition to the nuclear fuel cycle risks discussed so far, those generated by the production and handling of nuclear wastes certainly are among the most awesome of any risks known to humanity. If this critical responsibility could be properly discharged by the U.S. government, many who question nuclear power might be favorably impressed. But instead, a chilling and discouraging precedent has been set . . . .

NOTES

1. John W. Gofman, "The Cancer Hazard from Inhaled Plutonium," CNR Report 1975-1R (Dublin, California: Committee for Nuclear Responsibility, May, 1975). The Committee's address is P.O. Box 2329, Dublin, California 94566.

2. John W. Gofman, "Estimated Production of Human Lung Cancers by Plutonium From Worldwide Fallout," CNR Report 1975-2 (Dublin, California: Committee for Nuclear Responsibility, July, 1975).

---

*The two projects are a $662 million Fast Flux Test Facility at Hanford, Washington for the testing of breeder fuels, and a $1.8 billion demonstration breeder reactor to be built on the Clinch River near Oak Ridge, Tennessee. Costs have probably escalated since these lines were written.

3. Bernard L. Cohen, "The Hazards in Plutonium Dispersal," Report of The Institute for Energy Analysis (Oak Ridge Associated Universities, P.O. Box 117, Oak Ridge, Tennessee 37830, March 1975).

4. British Medical Research Council Committee on Protection Against Ionizing Radiations, "The Toxicity of Plutonium," British Medical Research Council (London, England: Her Majesty's Stationery Office, 1975).

5. Advisory Committee on the Biological Effects of Ionizing Radiation, "The Effects of Populations of Exposure to Low Levels of Ionizing Radiation," Division of Medical Sciences, National Academy of Sciences (Washington, D.C.: National Research Council, November, 1972).

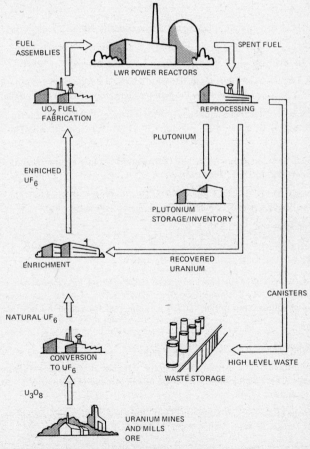

Figure 5-2. Chemical and metallurigical operations in nuclear fuel cycle for LWRs. (Note: no reprocessing plants are currently operating)

6. International Commission on Radiological Protection, Publication 19, "The Metabolism of Compounds of Plutonium and Other Actinides," adopted by the Commission (Oxford, England: Pergamon Press, May, 1972).

7. John W. Gofman, "The Cancer Hazard from Inhaled Plutonium."

8. W. J. Bair and R. A. Thompson, "Plutonium: Biomedical Research," *Science,* Vol. 183, pp. 715-722, 1974.

9. Arthur R. Tamplin and Thomas B. Cochran, "Radiation Standards for Hot Particles: A Report on the Inadequacy of Existing Radiation Protection Standards Related to Internal Exposure of Man to Insoluble Particles of Plutonium and Other Alpha-Emitting Hot Particles," February, 1974. National Resources Defense Council, 1710 N Street, N.W., Washington, D.C. 20036.

10. Quoted in Gofman, "Estimated Production of Human Lung Cancers . . . " July, 1975.

11. Jack Sheperd, "The Radiant Nuclei," *Intellectual Digest,* March 1973.

12. "Comparison of Fuels Used in Power Plants," Background Information, Public Affairs and Information Program, Atomic Industrial Forum, undated, p.2.

13. Thomas C. Hollocher and James J. MacKenzie, "Radiation Hazards from the Misuse of Uranium Mill Tailings," in D. F. Ford, *et. al., The Nuclear Fuel Cycle* (San Francisco: Friends of the Earth, 1974).

14. Arell S. Schurgin and Thomas C. Hollocher, "Lung Cancer Among Uranium Workers," in D. F. Ford, *op. cit.*

15. Robert Pohl, "Nuclear Energy: The Health Effects of Thorium-230" (Ithaca, New York: Cornell University, Department of Physics).

16. Hollocher, *op. cit.,* p.97.)

17. "Through the Mill With Nowhere To Go: NRDC Requests EIS on Uranium Mills," *Not Man Apart,* Vol. 5, No. 14.)

18. *Environmental Radiation Protection Requirements for Normal Operations of Activities in the Uranium Fuel Cycle.* Draft Environmental Statement for Proposed Rulemaking Action, U.S. Environmental Protection Agency, May, 1975.

19. *Ibid.*

20. *Congressional Record,* S. 19117-8, October 12, 1973.

21. J. Gustave Speth, Arthur R. Tamplin, Thomas B. Cochran, "Plutonium Recycle: The Fateful Step," *Bulletin of the Atomic Scientists,* Vol. 30, No. 9 (November, 1974).

22. Union of Concerned Scientists Staff, "Nuclear Fuel Reprocessing: Radiological Impact of West Valley Plant," in D. F. Ford, *et. al., op. cit.*

23. Thomas B. Cochran and J. G. Speth, *NRDC Comments on WASH 1327 Draft Generic Environmental Impact Statement on Mixed Oxide Fuels,* Na-

tional Resources Defense Council, Inc., 1710 N. Street, N.W., Washington, D.C. 20036.

24. Amanda Spake, "South Carolina's Silent Death Factory," *New Times* (January 24, 1975).

25. Spake, *op. cit.*

26. Robert Gillette, "Plutonium (I): Questions of Health in New Industry," *Science,* Vol. 185 (September 20, 1974).

27. *Ibid.*

28. *Ibid.*

29. Howard Cohen, "Malignant Giant," *Rolling Stone* (March 27, 1975).

30. "Theory in Death of a Critic Disputed," *New York Times,* as reprinted in the San Francisco Chronicle, December 24, 1974.

31. "Atom Worker Death Inquiry Disputed," *New York Times,* January 22, 1975.

32. "Death of Plutonium Worker Questioned by Union Official," *New York Times,* November 19, 1974.

33. "AEC Finds Evidence Supporting Charges of Health Hazards at Plutonium Processing Plant in Oklahoma," *New York Times,* January 8, 1975.

34. *Energy Finance Week,* Vol. 1, No. 12 (July 23, 1975).

35. "Administration Cuts Back Its Plan to Develop a Plutonium Reactor to Generate Electricity," *Wall Street Journal,* June 2, 1975.

36. J. Gustabe Speth, *et. al., op. cit.*

37. John McPhee, *The Curve of Binding Energy* (New York: Farrar, Straus, Giroux, 1974).

38. *New York Times,* January 8, 1975.

39. Allen Hammond, William D. Metz, Thomas H. Maugh, II, *Energy and the Future* (Washington, D.C.: American Association for the Advancement of Science, 1973), pp. 33-34.

40. Letter from Charles L. Hyder, physicist associated with Southwest Research and Information Center, Albuquerque, New Mexico, undated.

41. "Morton and Zarb Join in Suggesting a Slowdown on Nuclear Breeder Reactors and Call for More Research," *New York Times,* June 10, 1975, p. 25.

42. Thomas B. Cochran, *The Liquid Metal Fast Breeder Reactor* (Baltimore, Maryland: Resources for the Future, 1974).

43. Sheldon Novick, "Nuclear Breeders," *Environment,* Vol. 16, No. 6 (July/August 1974).

Dennis Renault-Sacramento Bee

"Someday, son, this will all be yours. And your son's. And your son's son's. And your son's son's son's. And his son's. And his son's son's . . ."

6

RADIOACTIVE GARBAGE

For the nuclear advocates, radioactive waste management is one of the most difficult problems of the nuclear era. It was a bane to the AEC and no doubt will plague the Energy Research and Development Administration and the Nuclear Regulatory Commission, for several reasons. First, commercial nuclear reactors produce large amounts of radioactivity. AEC planners have forecast a cumulative total equivalent to 60 million gallons of high-level waste by the year 2000. Rodger Strelow, an official of the U.S. Environmental Protection Agency, has stated that by 2000 A.D., federal and commercial nuclear power combined may generate up to 200 million gallons of high-level waste, 400 million cubic feet of low-level waste and 95 million cubic feet of alpha waste. No satisfactory solution has been found for disposing of this waste nor for isolating it perfectly from the environment Yet because of the large amounts of radioactivity involved and its potential ill effects, virtually perfect isolation is necessary.

HIGH-LEVEL WASTES

Many different kinds of radioactive elements are found in the high-level wastes produced by reactors. Some of these elements remain radioactive for periods of seconds and others for up to a hundred million years. Among the many formidable poisons created, several are of special interest for their biological hazards, particularly strontium-90, cesium-137, and plutonium-239. All are long-lived radioactive products, but, of course, short-lived elements, such as iodine-131, with an eight-day half-life, can also be health threats.

Strontium-90 and cesium-137, which are found under normal operating conditions in fissioning fuel rods, both emit extremely intense penetrating radiation and large amounts of heat as they decay. Both elements are generally considered to require guardianship and isolation from the environment for about a thousand years.

Strontium-90 is a bone-seeking isotope accumulated by the body and not excreted like many other isotopes. Just one gallon of high-level liquid waste may contain 50 to 100 curies of strontium 90. To dilute that much strontium-90 to meet legal guidelines for water purity would require 500 billion to a trillion gallons of water.

Like strontium, cesium is a potent carcinogen. It emits gamma rays, similar to x-rays, which can penetrate thick shields. If cesium is absorbed by the body, it concentrates in the muscle tissue, and in the ova of females. Unlike either strontium or cesium, the plutonium in nuclear wastes has low penetrating power. It must be inhaled, ingested, or implanted in the body for its phenomenally toxic effects to be induced.

Plutonium is a synthetic element that did not even exist 35 years ago until created by scientists working on the atomic bomb. As explained in more detail in Chapter 5, plutonium is the most lethal carcinogen known and single particles of it in oxide form (PuO_2) can produce lung cancer when inhaled. The AEC, the nation's largest plutonium producer, in 1972 projected that about 10 million pounds of plutonium will be produced commercially in the U.S. by the year 2000. One pound, if administered in finely dispersed particles to human lungs, is more than enough to give the entire human race cancer. ERDA publicly favors the use of reprocessed plutonium to supplement uranium fuel in conventional reactors and breeders. This would bring plutonium into widespread general commercial use.

Any mishap leading to the release of toxic plutonium oxide to the environment can be particularly grave because of plutonium's half-million-year guardianship period implied by the element's 24,400 year half-life. Small fragments of plutonium burn spontaneously on exposure to damp air, producing an almost invisible smoke. Scattered into air or onto soil, the plutonium can be suspended and resuspended to produce cancers long after the source and cause of its original dispersal may have been forgotten. Already, less than two generations after plutonium's creation, everyone on earth carries tiny trace quantities of it in their lungs, primarily from the nuclear weapons testing of the post-war period.

To deal with such toxic wastes as plutonium, strontium, cesium, and others, the AEC had planned to license at least three waste reprocessing plants by 1974 to recover unused uranium and plutonium from reactor wastes for use in making fresh reactor fuel. After extraction of the recyclable fuel components, the remaining wastes were to have been shipped to a final storage site. The three waste reprocessing plants were to have been the Nuclear Fuel Services plant at West Valley, New York; a General Electric plant at Morris, Illinois; and the Allied-General Nuclear Services plant near Aiken, South Carolina, in Barnwell County.* In reprocessing (discussed in Chapter 5), spent fuel rods are first dissolved in nitric acid and then subjected to a complex and extremely hazardous chemical process. As of early 1976, no commercial reprocessing plants were operating and none may ever operate. But under the provisions of various bilateral agreements between the U.S. and foreign nations, spent fuel from U.S. reactors sold abroad can be returned to U. S. facilities for extraction of resalable nuclear materials and perpetual storage of wastes. Now any such reprocessing will have to be done in government facilities until commercial facilities are available.

The G.E. Midwest Fuel Recovery plant in Illinois was supposed to open in 1970 at an estimated cost of $36 million.[1] But after more than six years of construction at a cost of $64 million, G.E. couldn't get the plant to work and has abandoned its plans to reprocess fuels there. Rebuilding the plant would have taken four more years and $90 to $130 million, with no guarantee of success, so it is understandable that the project has been scrapped. To mollify the facility's intended customers, G.E. is offering interim storage for their spent fuel at the plant site. "Some government

---

*The plant is a joint venture of Allied Chemical Corp. and General Atomic Co., a subsidiary of Gulf Oil Corp., and the Royal Dutch-Shell Group.

authorities," said Robert Gillette in *Science* magazine, ". . . see in G.E.'s predicament a critical lesson for the energy industry as a whole; that the perils of pushing new technologies too fast are great and costly." [2]

As mentioned in Chapter 5, the Nuclear Fuel Services plant is currently closed and Getty Oil has reportedly made unsuccessful efforts to sell it. [3]

The Barnwell plant, which was to have been completed in 1974 at a cost of less than $100 million, has now cost more than $250 million, and its start-up is likely to be delayed at least until the early 1980s, according to industry sources of the *Wall Street Journal*. Moreover, this plant cannot conduct sustained operations without two new auxiliary plants which would cost *at least* twice the total spent so far on the entire original plant.

One of these additional plants would solidify the stream of plutonium-bearing liquids from the Barnwell plant so that the plutonium could be shipped in solid form for conversion into fuel. The solidification facility was mandated in 1974 when the AEC banned the shipment of liquid plutonium, effective 1978.

The other auxiliary plant the Barnwell complex needs would convert Barnwell's high-level radioactive wastes from a liquid to a solid state.

Allied-General is balking at building both new processing plants, claiming the technologies the plants would employ have not yet been fully developed.[4] In addition, approval of plutonium recycling by the U. S. Nuclear Regulatory Commission would be necessary before plutonium fuel could be used commercially to fuel reactors, and the Commission still has not made a final decision on whether to allow this. If the approval is given, this would be tantamount to the inauguration of a plutonium energy economy in the U.S.

Faced with skyrocketing costs and with uncertainty about plutonium recycling, the ultimate justification for the plant, Allied-General is now asking for a huge government bailout. The company has proposed that the Energy Research and Development Administration should build the two additional plants. Then if the plants work and are licensed, Allied-General might take them over from the government.

The U.S. nuclear industry's failure to develop a viable reprocessing technology highlights the industry's tendency to under-

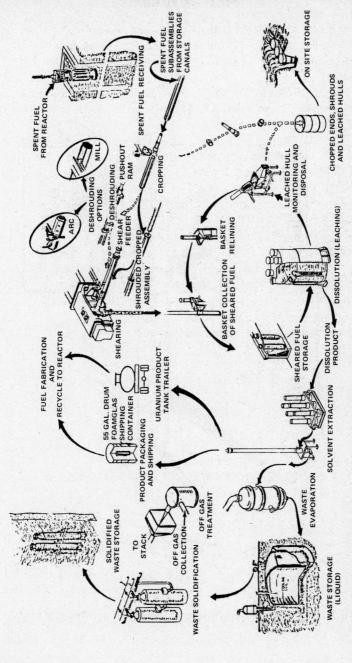

Figure 6-1. Reprocessing of Spent Power Reactor Fuel.

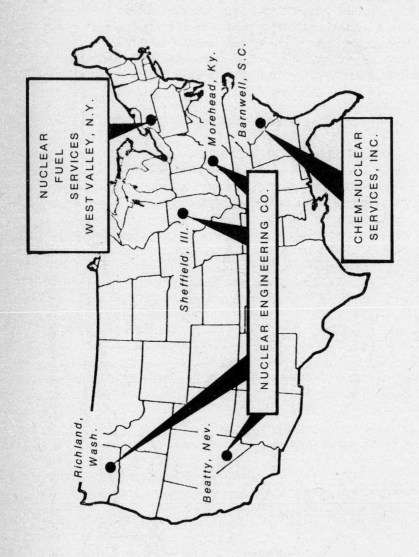

Figure 6-2. Commercial burial sites for low-level solid waste in U.S.

estimate the technical difficulties and the environmental problems that a headlong rush to a nuclear economy entail. An identical federal regulatory philosophy permits nuclear plants to operate with unproven Emergency Core Cooling Systems allowing hazardous wastes to pile up at nuclear power plants and in makeshift interim storage sites before satisfactory means are available for reprocessing the wastes or safely disposing of them.

INTERIM STORAGE SITES

Currently, ERDA has the responsibility for nuclear wastes accumulated by the AEC at the Idaho National Engineering Laboratory, formerly the AEC's National Reactor Testing Station near Idaho Falls, Idaho. There, the AEC established a radioactive waste storage area about six hundred feet above the Snake Plain Aquifer. This huge underground river connects to the water supply for much of the Pacific Northwest. If storage devices fail, contamination of the aquifer might eventually ensue.[5]

A second federal storage site, euphemistically called a "tank farm" by the AEC, is located near Aiken, South Carolina, at the "Savannah River Reservation." Millions of gallons of high-level waste sit in aging tanks there; AEC planners proposed pumping them beneath the Tuscaloosa Aquifer, a major source of water for the public and for industry, but the idea was not well received. An AEC advisory committee's report which the AEC suppressed for four years termed the plan "in its essence dangerous."

About 75 percent of the nation's nuclear wastes are stored at ERDA's Hanford Reservation on the banks of the Columbia River in southeastern Washington, near Richland. Concentrated wastes so hot they boil from their own radioactivity are kept in gigantic concrete-clad steel tanks which have a usable life of about twenty to thirty years. Some hold up to a million gallons of liquid. A number of tanks are water cooled; others allow the liquids to boil off, leaving a moist but solid residue. (Vapors are condensed and scrubbed to prevent the escape of radioactivity.)

The federal government took over the four hundred thousand-acre Hanford Reservation in February 1943 in order to produce plutonium for the nation's atomic bomb project, conducted by the Manhattan Engineer District. Eventually, a Purex nuclear ma-

terials processing plant and nine reactors were built at Hanford, all
but one now retired. Reactor fuel was irradiated in the reactors and
the resulting plutonium was used in the Nagasaki bomb, or to fill
U.S. nuclear stockpiles with atomic weapons. Some plutonium
was solidified and sent to the AEC's $250 million Rocky Flats
Nuclear Weapons Facility near Denver to make triggers for hydrogen bombs.*

During World War II and for the following twenty-five years,
about 70 million gallons of high-level wastes from these activities
accumulated at Hanford. In 1973, about 25 million gallons had
been reduced to a supremely radioactive moist salt sludge and
about 45 million remained as liquid, much of it in tanks between
twenty and thirty years old.† The General Accounting Office (which
extends investigative aid to Congress, among other functions), the
U.S. Geological Survey, and private consultants all issued warnings
to the AEC from 1953 to 1971 about the precarious condition of its
tank storage, but the AEC was reluctant to make the investments
necessary to correct the situation. Consequently, the warnings went
largely unheeded, and, to date, a total of more than half a million
gallons has leaked from Hanford's tanks. Large leaks have also
sprung in transmission lines carrying wastes from plant to tanks.

Just one 115,000-gallon leak of high-level waste at Hanford
spilled an estimated 14,000 curies of strontium, 40,000 curies of
cesium-137, and 4 curies of plutonium into the soil. It went unnoticed for nearly two months and, when it was finally discovered,
emergency pumping of the wastes remaining in the leaky tank did
not begin until more than a day later. A report by an AEC team
analyzing the incident said that worn-out tanks and primitive monitoring contributed to the spill, but that its primary cause was negligence by the Atlantic Richfield Hanford Company, which super-

---

*For a discussion of the fires and accidents that have released plutonium to the
general environment beyond this plant's boundaries, see, "The Plutonium Situation at Rocky Flats," produced by the Nuclear Information Center, 2239 East
Colfax, Denver, Colo. 80206, November 1975 ($.50).

†In addition to the wastes at Hanford, about 18 million gallons of high level
wastes are stored at ERDA's Savannah River Plant near Aiken, South Carolina,
and about 2 million gallons are stored at the agency's Idaho National Engineering
Laboratory. Most of the wastes so far have been produced by the AEC in weapons
production, research and development, and nuclear submarine fuel reprocessing
activities.

vises the day-to-day operation of nuclear waste storage for the AEC at Hanford.

In a masterly journalistic account of the accident for *Science*, Robert Gillette wrote:

> More to the point is what the incident reveals about the keenness of the AEC's vigilance over the nation's vast and expanding store of nuclear processing wastes. . . . Is the AEC really prepared to manage thousands of pounds of wastes that civilian nuclear power plants will be generating in the years ahead? And how, exactly, could it lose the equivalent of a railroad tank car full of radioactive liquid hot enough to boil itself for years on end and knock a Geiger counter off scale at a hundred paces?[6]

The AEC's post-mortem investigation of the huge leak revealed poor management at the nation's chief waste repository. Training, communications, and supervisory failings of the most serious nature were found.

The history of accidents at Hanford reveals that some of the waste tanks had been sporadically leaking thousands of gallons since 1958, despite a reassurance in 1959 by the manager of the Hanford chemical plants that no leaks had been discovered. (At the time of that disclaimer, two leaks already were on record.) In addition, the AEC from 1963 to 1965 was without spare tanks to use in emergencies. When a leak developed in tank 105-A in November, 1963, plant operators had to wait until it sealed itself. They allowed the tank to remain in use and subsequently filled it fuller than before — 10 percent over its design capacity.

A 1968 General Accounting Office report which found numerous defects in the AEC waste management program was classified and kept from the public until 1970. The earliest warning, a U. S. Geological Survey report on groundwater at Hanford, was kept classified until 1960 and was not published until 1973. A National Academy of Sciences study concluding that AEC waste management practices were unsafe was suppressed from 1966 to 1970, until Senator Frank Church of Idaho compelled its release.*

Since 1961, the AEC has built new tanks and has retired some of

---

*That document entitled *Report to the Division of Reactor Development and Technology* (USAEC, May 1966) was prepared by the National Academy of Sciences' Committee on Geological Aspects of Radioactive Waste Disposal, which acted as a scientific advisory group to the AEC.

its more decrepit models. Tank monitoring at Hanford is better now, enabling ERDA to catch leaks quicker — but not prevent them entirely.*ERDA is eager for commercial waste reprocessing to become a reality, but even this will not end the waste management problems at Hanford.

To make leakage from the Hanford tanks more difficult, the AEC/ERDA began removing some radionuclides from high-level liquid wastes in some of its tanks, and is solidifying the remaining liquid into a moist salt cake in those tanks. But current technology is incapable of completely evaporating all the fluid from the wastes, and an extremely concentrated highly caustic fluid known as "terminal liquor" remains. The liquor is much more corrosive than the original more dilute liquid wastes.

Consequently, even after solidification to moist salt, high-level wastes will continue corroding their tank walls at Hanford. Moreover, the resulting deposits are rapidly becoming difficult to remove from their tanks. In a memorandum on ERDA's high-level waste storage program, the Natural Resources Defense Council ,stated, " . . . *within a few years ERDA may be unable to retrieve the wastes economically and safely"* (original emphasis).

Fifteen tanks are already so badly decayed that their waste sludge can no longer be retrieved with present technology. Removal requires flushing the tanks with high pressure water to resuspend and pump out the wastes. This would probably dissolve sludges now plugging leaky tank areas and would result in major leaks. Yet as of mid-1975, ERDA still had no acceptable alternative but to keep its high-level wastes in hazardous tank storage.

In addition to its tank storage "farms," ERDA also uses trench waste storage. The AEC deliberately put a great deal of plutonium-bearing high-level "soup" directly into the ground at Hanford in earthen trenches with concrete tops and sides. Plutonium is insoluble in alkaline water and, because the groundwater at Hanford is alkaline, the AEC's waste disposal theorists reasoned that the plutonium would not dissolve and would remain in the soil

---

*In April, 1975, ARCO-Hanford officials reported tank 104-A released 5,000 gallons of high-level liquid containing strontium, cesium, and ruthenium from the bottom of the container. A company spokesman said he thought the liquid would not travel more than 10 feet from the bottom of the tank, and that the water table was 233 feet below.[7]

indefinitely. The National Academy of Science Committee on Geological Aspects of Radioactive Waste Disposal disagreed:

> The current practices of disposing of intermediate-level and low-level liquid wastes and all manner of solid wastes directly into the ground above or in the fresh-water zones, although momentarily safe, will lead in the long run to a serious fouling of man's environment."

Thomas A. Nemzek, then general manager at Hanford, has claimed that no high-level waste ever reached groundwater at Hanford and that, even if it would, the Columbia River would still remain within safe drinking standards. Recent evidence, however, casts doubt upon the AEC's studies of the Hanford area's geohydrology and suggest that plutonium may be more mobile in soils than previously thought.[8]

The Columbia River's purity is not a new subject of controversy. In May 1964, the Federal Water Pollution Control Agency said Hanford reactors were contaminating fish and oysters with phosphorus-32 and zinc-65. The Agency in 1966 charged that the AEC's negligence "has resulted in the worldwide acknowledgment of the Columbia River as the 'most radioactive river in the world.'" The suppressed NAS report on waste, released by Senator Church and noted earlier, observed that seepage of the radioactive isotopes tritium and ruthenium might eventually contaminate the Columbia.

But the impact of radioactive wastes spilled at Hanford can extend far beyond even the great Columbia River. Radionuclides can spread rapidly in solution from the mouth of the Columbia, through the North Pacific Ocean as far as Japan. By biological transfer and concentration processes, the radioactivity will ultimately taint edible fishes such as albacore that end up on dining room tables. As Sierra Club Research Director Dr. Robert R. Curry noted in the club's October 1974 *Bulletin*, Hanford wastes "pose a serious threat to food chains of the whole Pacific."

The discharge of plutonium directly into the ground trenches entailed other serious risks, and the practice was stopped in late 1973. The AEC previously had placed such a huge quantity of nuclear waste in a trench, referred to as Z-9, that some scientists feared the mass of plutonium there could become critical and that an explosion might occur as a result of trench flooding after a heavy snow and thaw. To avoid a criticality incident, the AEC "poisoned" the trench with neutron-absorbing cadmium nitrate, but publicly acknowledged the dangerous situation it had created. At one stage in its handling of the Z-9 affair, the AEC asked Congress for $1.9 million to mine the plutonium out of Z-9 while one and one-half million gallons of plutonium were being put into trench Z-18.

Another grave danger, noted by an Environmental Protection Agency official who participated in the squelched NAS study, would be a change in the water at Hanford from alkaline to acidic. This would greatly increase the solubility of plutonium and, consequently, its transport into ground water systems. Crop irrigation runoff, industrial wastes, or even sewage from subdivisions are all capable of adding acidifying chemicals to soil. During a period of half a million years, the isolation period for plutonium, some such unforeseen changes might alter the water chemistry.

During the last decade alone, some of the waste disposal notions considered safe and foolproof by AEC waste management researchers have now come to be regarded as too dangerous by many distinguished scientists. Maybe ERDA researchers today are still making similar unperceived potentially fatal errors in judgment with respect to other areas of the nuclear power program.

Within the next million years, it is conceivable that major climactic changes might occur, flooding large areas of the U.S. that are now on its coasts. Other scientists expect that within a few thousand years, a new Ice Age may once again send glaciers scouring over the land from the north. The thought of a glacier ripping open a plutonium crypt may seem bizarre and far-fetched, but the strangeness of that image is another indication that U.S. nuclear planners have unwisely flaunted nature by creating enormous amounts of radioactivity without possessing the knowledge to handle it.

The term "waste disposal" as used in reference to radioactive materials appears to be a misnomer today, in that their "disposal" is currently beyond our technological capacity, and the best that ERDA can do is to store wastes until a disposal method can be proven. ERDA tacitly admits that no satisfactory solution even to the problem of waste management has been found. As recently as February 1976, ERDA's chief administrator, Dr. Robert C. Seamans, said the agency *intends* to make a national survey of potentially suitable underground waste storage sites. The survey will consider a variety of geological formations: salt beds, shales, crystalline rock, such as granite, and volcanic deposits. Perhaps this kind of research should have been done before the nuclear power program was rushed into a full-blown commercial industry by intense government promotion and expensive federal subsidies.

The notion of burying nuclear wastes far underground in salt beds or domes, or in solid rock, is viewed by ERDA officials as the most promising long-term storage method of all. Consequently, ERDA is currently investigating the use of an underground salt bed about thirty miles east of Carlsbad, New Mexico, as the country's first pilot underground waste storage facility. Before wastes are finally buried in caverns hollowed out of the underground salt

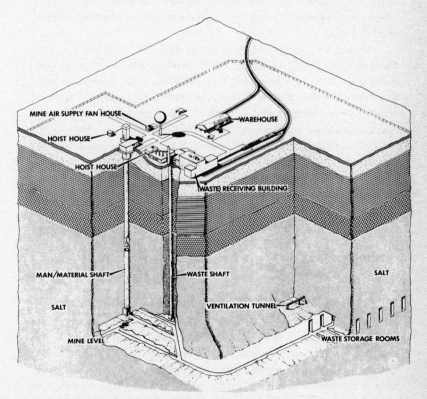

Figure 6-3. A Proposed pilot Bedded Salt Waste Storage Plant

beds, ERDA plans to develop acceptable technology for the solidification of liquid wastes into glassy or ceramic-like solids.

Several years ago, the AEC was preparing to use a salt mine at
Lyons, Kansas, for this purpose until the site was shown to be
unsuitable by the Kansas Geological Survey. ERDA has now commissioned Sandia Laboratories of Albuquerque, New Mexico, to
study a site called Los Medanos near Carlsbad, New Mexico, for
use as a disposal area. In the salt bed storage method, the wastes
would be put in a cavity and cooled by pumped water for several
years. Then the cavity's entry shaft would be plugged, the water
would boil away, and in a few decades, the heat would seal the
wastes by fusing surrounding rock while the heat kept water away.
It would also melt the waste burial canisters so that — if anything
went wrong — retrieval of the wastes would be extremely difficult
or impossible.

Nonetheless, salt formations are preferred by disposal theorists;
the presence of solid salt, with its great solubility in water, indicates that salt bed and dome sites have been free of water for a
long time. Whether they will remain so in the future is another
question, especially in view of man's many mining and well-drilling
operations. More questions remain to be answered about this disposal
method and Batelle researchers have predicted geological storage
will not be available for fifteen to thirty-five years.

Perhaps discouraged by the risks, incompetence, and failures
of the waste management program, scientists have created a new
science fiction, rife with futuristic disposal schemes. These include
proposals to bury wastes under polar ice caps, under the ocean
floor, in deep holes, or in caverns blasted in rock. One proposal
even calls for the use of atomic bombs to create disposal caverns
underground near reprocessing facilities. Another suggests rocketing
the wastes into the sun. None of these solutions has been tested
under anything approaching actual conditions, and most are completely untested. Some are technologically impossible; others are
phenomenally costly. Almost all these schemes depend on the complete cooperation of the biosphere: The earth's crust must agree
not to fold, thrust, quake, lift, or break in any area containing
long-lived wastes. The oceans must abide in their present beds, and
certain large-scale climatic aberrations must not be allowed to occur.

Scientists have proposed that burying radioactive wastes in polar ice caps would dissipate the wastes' heat and allegedly keep the wastes isolated. But what if the material melted the ice more quickly than expected and found its way into the oceanic food chains upon which the survival of life on the planet depends? What if the polar caps melted from climatic changes sooner than expected?

Another suggestion is to bury wastes in holes five to seven miles beneath the oceans' surface by drilling into the ocean bedrock. This method is predicated on the assumption that such areas are unlikely to be mined. Fifty years ago similar assumptions might have been made about the seabeds of the outer continental shelf where oil drilling now occurs.

Other theorists have suggested digging the holes in the ocean in locations where buried wastes might slide underneath great land masses as a result of "continental drift." That theory says that continents are not immovably anchored on the earth, but can shift about on a relatively yielding, underlying surface. When one layer of the earth's crust inches forward, the theory states, another layer is drawn beneath it or "subducted." No one knows precisely where the material would have to be buried for it to be subducted, and the postulated continental drift may be exceedingly slow and slight.

Use of the oceans as a dumping ground is not new. From 1946 to 1970, the AEC simply dumped hot wastes onto the ocean floor in concrete cylinders that sank and soon began to leak in the corrosive salt water. The U.S. dumped an estimated 97,000 curies of waste during this period at two locations. One of the dump sites was about forty miles offshore from San Francisco near the Farallon Islands. The other was one hundred twenty miles from the Maryland-Delaware coast. Several European nations have dumped an estimated 240,000 curies into the North Atlantic.

In a controversial study for the AEC by Batelle's Pacific Northwest Laboratory, Batelle estimated that the cost of seabed waste disposal might be as much as $5 billion. It is now widely believed that this figure is an overestimate, but it is clear that waste storage is going to be expensive. The EPA's Roger Strelow has quoted the cost of waste storage for the next twenty-five years at about $8 billion.

Figures like these have led critics of the nuclear industry to state, "Production of nuclear energy is exceedingly wasteful and it is only profitable due to monopoly. But if suppliers were forced to develop and maintain really safe systems of waste disposal, then the profit would disappear entirely."[9] Even for $5 billion, seabed disposal schemes are not about to solve our nuclear waste problems. Batelle has estimated it might be 1995 or 2000 before research on it will be completed. Another "solution" is shooting nuclear waste into the sun. Apart from the tremendous cost of sending up as many as nine rockets a day by the year 2000 to ferry out the garbage, other formidable problems remain. What would happen if a rocket exploded on the launching pad, scattering its nuclear payload? What if a rocket misfired and crashed back to earth, dispersing a hideous cargo?

Do we have the right to dissolve our poisons in the oceans or leave buried poison caches in ice or in the ground, where they may blight the environment for future generations and despoil the earth for thousands of years?

With no safe long-term storage facilities available and no reprocessing, ERDA has been obliged to resort to hazardous and expensive interim-storage, requiring constant monitoring. In short, the AEC encouraged, and ERDA acquiesced in, the production of huge stockpiles of lethal wastes although no proven method is available for long-term storage. Mysteriously, we seem to have come to accept without question the great subsidies the AEC and ERDA have bestowed on nuclear power companies by doing their waste disposal research. Why did the AEC foster commercial nuclear power and encourage national reliance on it without requiring that *the industry* first develop a safe and socially acceptable disposal technology?

Apparently stymied by the nuclear waste disposal problem, the AEC has attempted to ignore it. In a major draft environmental impact statement issued in 1974,[10] AEC nuclear waste experts contended that no permanent storage site would be needed for ten to a hundred years. Instead, the AEC experts recommended that wastes be stored on the surface of the earth in specially designed containers until a better solution can be devised. Thus the AEC avoided having to consider the environmental impacts of waste disposal. Dr. Terry R. Lash, a staff scientist for the Natural Resources Defense Council observed, "The AEC has improperly substituted consideration of an extremely short-term and more manageable problem for the real one. . . . "[11]

For example, the AEC once advocated encasing wastes in canisters within steel casks inside 55-ton concrete pillars above ground. The project was known as the Retrievable Surface Storage Facility, and its attraction was that, in contrast to some underground storage proposals, if anything went wrong with a waste container, its contents could be "retrieved." These waste casks were to have been kept immersed in water or cooled by constantly circulating air, and they would have demanded careful surveillance. But major questions arose about the longevity of these concrete monoliths, their maintenance costs, and how a leaky container at one of these Stonehenges would be repaired or serviced. Could their radioactivity then be adequately contained? The project was quietly dropped.

LOW-LEVEL WASTES

Because low-level radioactive wastes do not seem to present as immediate a potent health hazard as do high-level wastes, their management has been neglected by the federal government and by some commercial waste-storage-facility operators. Low-level wastes have been buried in shallow earthen trenches in six U.S. states in containers that will disintegrate long before the wastes they contain lose their toxicity. Perhaps this neglect of low-level wastes is in part due to the misleading application of the term "low-level" to designate wastes that may contain plutonium and americium and other transuranic elements. To the uninitiated, "low-level" may connote "innocuous." But though low-level wastes require little shielding and have negligible heat-generation rates, some scientists believe that the amounts of plutonium in low-level wastes may be even greater than the amounts in high-level wastes. [12]

Practically unknown to the general public, radiation is already escaping from low-level waste burial sites. This has been documented in the cases of burial trenches at West Valley, New York and Maxey Flats, Kentucky. Samples of tritium, cobalt-60, cesium-134, cesium-137, plutonium-238, plutonium-239, strontium-89, and strontium-90 have been released in Kentucky. It is likely that migration of radioactivity from other commercial burial sites has already occurred or is imminent. The absence of further evidence of such leakage probably stems from the fact that few

Figure 6-4. Typical burial operations at a federally licensed burial ground for low-level solid wastes

studies have been done on these sites. It appears that no complete safety analysis has ever been completed for any commercial low-level waste burial site.

CONCLUSION

The gaps that exist today in our highly experimental waste management technology have 'been a source of great concern to many scientists and lay people. A special task force set up by the 1973 Pugwash Conference on World Affairs evaluated the nuclear wastes problem and concluded:

> With respect to the management of long-lived radioactive wastes, strong uncertainties still exist. The principal difficulty is that the material remains highly toxic for periods measured in thousands of years; even over shorter spans, predictions about the stability and continuity of human society are impossible, and over the longer term significant geological change is possible in some circumstances. . . .

These words of caution by some of the world's foremost scientists, combined with the commonly available knowledge that the AEC has on numerous occasions grossly mismanaged nuclear wastes already, leads to the inescapable conclusion that the nuclear industry should not generate nuclear risks which they are not ready to handle.

NOTES

1. "Radioactive Wastes: The AEC's Non-Solution," Natural Resources Defense Council, reprint from NRDC Newsletter, Winter, 1974-1975.

2. Robert Gillette, *Science,* August 30, 1974.

3. Thomas B. Cochran and J. G. Speth, *NRDC Comments on WASH 1327 Draft Generic Environmental Impact Statement on Mixed Oxide Fuels,* Natural Resources Defense Council, Inc., 1710 N Street, N.W., Washington, D.C. 20036.

4. "White Elephant? Big Plant to Recycle Nuclear Fuel is Hit By Delays, Cost Rises," *Wall Street Journal,* February 17, 1976.

5. "Atom-Age Trash, Finding Places to Put Nuclear Waste Proves a Frightful Problem," *Wall Street Journal,* January 25, 1971.

6. Robert Gillette, "Radiation Spill at Hanford: The Anatomy of an Accident," *Science,* August 24, 1973.

7. "Hanford Comes Through Again," *Not Man Apart,* Vol. 5, No. 11 (June, 1975).

8. Natural Resources Defense Council, Inc., Supplementary Statement to Hearing Board on AEC's draft programmatic environmental impact statement, *Waste Management Operations, Hanford Reservation,* Richland, Washington (WASH-1538), February 6, 1975.

9. Ernest Winter, a consultant to the U.N. Environment Program, quoted in "How to Handle Nuclear Waste," *San Francisco Chronicle,* April 12, 1974.

10. *Management of Commercial High-Level and Transuranium-Contaminated Radioactive Wastes.* WASH-1539, Draft, U.S. Atomic Energy Commission, Washington, D.C.

11. "Radioactive Wastes: The AEC's Non-Solution," Natural Resources Defense Council, Winter, 1974-1975.

12. U.S. Environmental Protection Agency, "Comments (0-AEC-A00107-00) on Management of Commercial High-Level and Transuranium-Contaminated Radioactive Wastes (WASH-1539)," p. 11 (November 1974); and Papadoulos and Winograd, U.S. Geological Survey, "Storage of Low-Level Radioactive Wastes In The Ground Hydrogeologic and Hydrochemical Factors With An Appendix On The Maxey Flats, Kentucky, Radioactive Waste Storage Site: Current Knowledge and Data Needs For A Quantitative Hydrogeologic Evaluation," Open-File Report 74-344, 1974, as cited in Terry R. Lash, Address to Third Annual Illinois Energy Conference on Nuclear Power In Illinois, Session on Environmental Effects, Pick Congress Hotel, Chicago, Illinois, Sept. 12, 1975. Copies of Lash remarks are available from Natural Resources Defense Council, Inc., 664 Hamilton Avenue, Palo Alto, Calif. 94301. See also T. H. Pigford, "Radioactivity In Plutonium, Americium and Curium in Nuclear Reactor Fuel," (A Study for the Energy Policy Project of The Ford Foundation) p. 36 (June 1974), as cited in Terry R. Lash (with the assistance of John W. Gofman), "Comments of the Natural Resources Defense Council on the Environmental Protection Agency's Draft Environmental Statement ENVIRONMENTAL RADIATION PROTECTION REQUIREMENTS FOR NORMAL OPERATIONS OF ACTIVITIES IN THE URANIUM FUEL CYCLE And PART 190-ENVIRONMENTAL RADIATION PROTECTION STANDARDS FOR NUCLEAR POWER OPERATIONS," (Palo Alto: Natural Resources Defense Council, 1975).

7

THE COST SPIRAL
and the
URANIUM SHORTAGE

"The long-held dream that nuclear power would give the United States and the world an endless stream of low-cost electric power has faded, according to a growing number of economists, technical experts and utility officials."
— David Burnham, "Hope for Cheap Power From Atom Is Fading," *New York Times,* November 16, 1975

More than a trillion dollars is scheduled to be invested in nuclear power during the next twenty-five years[1] — a huge chunk of domestic capital resources — yet nuclear power today is not an economically viable technology. The U.S. nuclear industry was conceived in subsidy, is dedicated to subsidy, and, without subsidy, cannot continue its current headlong expansion. The industry would not even exist and cannot thrive without further massive federal gifts, even though more than a hundred billion dollars has already been invested in nuclear power[2] and 20 percent of the generating capacity of New England is already nuclear.[3] For the technology that nuclear proponents claimed could provide as much as 60 percent of our central station generation by the year 2000 has reached an economic dead-end.

The nuclear power industry has generated its gigantic economic momentum on two premises: that nuclear power would be cheap for the consumer and that, although subsidies might at first be necessary, eventually the industry would grow strong and economically self-sufficient. We are about to see how economically *unviable* the nuclear industry is, and why its costs are bound to become progressively more extravagant.

The industry's economic difficulties result from inherent structural problems. Despite industry projections, U.S. nuclear expansion has already been halted by severe economic and political difficulties. From thirty-eight reactors ordered by domestic utilities in 1973, the U.S. nuclear industry received only four new domestic orders in 1975 and, in the first three months of 1976 the entire industry has received *no new domestic orders at all.* Of the two hundred fifty nuclear plants that nuclear advocates forecast would be in operation in 1985, no more than fifty-five are currently operating and less than seventy are under construction. Unless the industry were suddenly invigorated, it is hard to see how there will be more than half the projected number of plants operating by 1985.

The utilities' enthusiasm for nuclear power is dimming — in part because of rapidly increasing amounts of capital required to construct the plants, the long delay periods in getting the plants "on line," and nuclear industry's poor performance at generating electricity. All these troubles portend an ignominious end to the

Table 7-1.

STATUS OF U.S. NUCLEAR POWER GENERATING UNITS AS OF DECEMBER 31, 1975

Reactors holding commercial operating licenses 56
Reactors operated by ERDA producing commercial power 2
Reactors for which construction permits have been granted 69
Reactors for which site work has been authorized 18
Reactors ordered 72
Projects announced for which reactors have not been ordered . . .21
Total existing and prospective reactors in survey 238
Reactors canceled in 1975 12
Reactors deferred in 1975 72

SOURCE: U. S. Energy Research and Development
Administration, 1976.

industry's expansion unless further major government subsidies are forthcoming. Many of these subsidies have been proposed by President Gerald Ford in his planned hundred-billion-dollar Energy Independence Authority, a public agency strongly advocated by Vice President Nelson Rockefeller. The President resubmitted it to Congress with his fiscal year 1977 budget.

CAPITAL COSTS

Nuclear power's cost relative to its nearest competitor, coal power, hinges on the interplay of capital cost factors (such as construction costs and interest) with fuel costs for uranium oxide and for coal.

Capital costs for nuclear plants are higher than for coal plants, but the fuel for a reactor today still costs less than the coal for a comparable coal plant. A utility that decides to buy nuclear instead of coal generating capacity resembles the car buyer who opts for a more expensive model because of anticipated long-run savings through better fuel economy. Nuclear utilities essentially took a gamble that their expected savings on uranium fuel over coal costs would make the extra capital spending on nuclear plants worthwhile.

Things did not work out that way. Nuclear capital costs in 1973 were 50 to 280 percent higher than the AEC had forecast.[4] In 1967, the AEC estimated that typical capital costs were about $134/kw of nuclear capacity. For large nuclear plants of 1000 Mw (a million kilowatts), the cost would therefore have been $134 million. But by 1976, 1000-megawatt nuclear plants slated for delivery in 1985 were expected to cost as much as $1,200/kW — $1.2 billion for the entire plant and its first fuel load. Meanwhile, the nuclear fuel costs that were supposed to compensate for expensive nuclear plant costs careened upward, too.

Even those who today make the controversial claim that nuclear power holds a cost advantage over coal have had to admit that the assumed advantage is rapidly eroding. For example, a study by Irwin C. Bupp of the Harvard Business School and three co-authors from MIT has shown that the average annual dollar increase in nuclear capital costs has been more than twice the average increase for coal plants during 1969 to 1975.[5] Moreover, the Harvard/MIT group found that nuclear power's capital costs are streaking ahead of coal's at the rate of $19/kw every year. Every five years, that would add another $95 million to the cost of the nuclear plant. The Harvard/MIT group wrote, "The capital costs of large light-water reactors show no signs of stabilizing and, indeed, are apparently still climbing at alarming rates. . . . The present trends in reactor capital costs are significantly narrowing the economic gap between the two technologies. It would be a great mistake to assume that the present economic advantage held by LWRs [light water reactors] is permanent."

The problem of raising large sums of capital to keep up with the galloping increases in nuclear capital costs has been aggravated by the weakened financial position of domestic electric utilities as a result of oil price increases and high interest rates.

The fourfold increase in imported oil prices from 1972 to 1974 brought higher utility prices to consumers and this helped trigger sudden drops in electricity-consumption growth rates.[6] While the average per capita annual increase in energy consumption was 3.7 percent from 1965-1975, per capita average energy consumption *fell* 2.8 percent in 1974 after the oil price escalations.[7] Kilowatt hour sales of electricity in the northeastern U.S. dropped nearly 7 percent in 1974,[8] and almost 40 percent of the total amount of electricity which U. S. utilities could have generated in 1974-1975 remained in reserve because of low demand.[9] (The Federal Power Commission recommended an 18 percent reserve.) As a result, utility earnings fell and investor confidence in them dropped, making it difficult and, in some cases, impossible for utilities to raise capital at interest rates acceptable to them.[10]

Noting the nearly tenfold increase in nuclear capital costs that has occurred in the last decade, Daniel F. Ford, executive director of the Union of Concerned Scientists, estimates that capital costs for nuclear plants starting up in the early 1980s may well be $1,500 to $2,000/kW.[11] This would virtually insure a cost advantage for coal plants. Ford expects nuclear cost escalation to continue because of current and anticipated safety-related plant design and construction regulations.

NUCLEAR FUEL COSTS

Although the overall cost of nuclear electricity today depends only slightly on the cost of uranium, *the cost differential* between nuclear power and coal power is quite sensitive to uranium price gains. A ten dollar per pound increase in uranium oxide leads to a tenth of a cent (one mill) per kilowatt-hour increase in the price of delivered nuclear electricity and even the Atomic Industrial Forum, a nuclear industry association, claims that nuclear power has only a 3.2 mil per kilowatt-hour advantage over coal.[12,13,14]* Clearly this narrow margin could be wiped out by a sudden thirty-dollar increase in uranium costs. So even the "cost advantage" claimed by nuclear power's most ardent defenders is at best marginal, and would quickly disappear if uranium fuel costs rise. (In the fuel-cost analysis that follows, we will not even consider increases in the costs of fabricating uranium oxide into pellet-filled fuel rods, nor the costs of shipping the rods and disposing of the wastes.) Although the sharp rise in uranium costs is likely to continue, experts in both the coal and nuclear industries expect the price of coal to increase moderately at only 5 percent per year.[15] We will see why a dramatic rise in uranium prices will throw the nuclear industry into a crisis that will dwarf all previous nuclear setbacks.

In the late 1960s, General Electric projected future uranium prices over the long term would be around $4 to $4.50 a pound.[16] But by early 1976, uranium for immediate delivery had reached $37 a pound, almost nine times greater than the GE estimate. And bids for 1980 uranium are already near $50 a pound, according to the Nuclear Exchange Corporation of Menlo Park, California.

It is an understatement to say that uranium price escalation has caught the nuclear industry unprepared. The country's major supplier of nuclear fuel, Westinghouse Corporation, misread the trends in uranium prices so badly that in September 1975, the company announced it would have to default on its uranium supply agreements to twenty utilities which had contracted with Westinghouse for uranium ore deliveries in the 1980s and, in some cases,

---

*See note 14 at end of this chapter for an explanation of the strongly biased method which the Atomic Industrial Forum used to obtain this misleading result.

into the 1990s. The supply cutoff will occur in 1978, the company said, because of the unprofitability of the uranium supply business due to "dramatic uranium price increases."

The company's uranium contracts called for delivery of 70 million pounds of uranium, much of it now at prices far below current or expected market prices. Westinghouse does not own all of the uranium it contracted to sell, so it would have to acquire it at high prices and sell it cheaply.

Westinghouse's actions have been a blow to the nuclear utilities and reactor sellers. One Westinghouse fuel customer, Union Electric Co., a large mid-western utility, has found itself obliged to agree to pay $40 a pound for a million pounds of uranium to be delivered in 1979, and that will be just shy of a year's fuel supply for the company's two 1,150 Mw reactors now under construction. While Union looks for more uranium, it has joined with at least twelve other utilities in suing Westinghouse to force fulfillment of the fuel contract.

Already some utility executives, like those of South Carolina Electricity and Gas Co., have seen the writing on the wall: In September 1975, they canceled their plans for their second nuclear power plant near Parr, South Carolina, because they could not find an "acceptable" supplier of uranium ore, and because of increasing nuclear construction costs.

The Westinghouse incident and a great many accounts in major newspapers during 1975 have signaled the existence of a critical uranium shortage in the U.S. today and on world markets. But the implications of this shortage are not yet understood by important segments of the nuclear industry, nor by the public.

The U.S. today has only 690,000 tons of known uranium reserves, according to the U.S. Energy Research and Development Administration.* The difference between reserves and resources is an essential one in that estimated U.S. resources are far in excess of the reserve figures. A reserve is a raw material deposit of known size and location having a mineral content rich enough so that the value of the mineral to be recovered will exceed the recovery costs, when all of those costs are tallied. (Therefore, prevailing market

---

*This figure is somewhat inflated by the inclusion of an estimated 90,000 tons of uranium technically obtainable over the next quarter century as byproducts of copper and phosphate processing.

Table 7-2.

Estimated U.S. Uranium Reserves and Potential Resources, December 31, 1975

Tons U_3O_8

Production cost	Reserves	Potential Probable	Potential Possible	Potential Speculative	Total
$10	315,000	440,000	420,000	145,000	1,320,000
$10-15 Increment	105,000	215,000	255,000	145,000	720,000
$15	420,000	655,000	675,000	290,000	2,040,000
$15-30 Increment	180,000	405,000	595,000	300,000	1,480,000
$30	600,000	1,060,000	1,270,000	590,000	3,520,000
By Product*					
1975-2000	90,000	-	-	-	90,000
2000-2020	150,000				150,000
	840,000	1,060,000	1,270,000	590,000	3,760,000

*By-product of phosphate and copper production. Source: US ERDA

prices for uranium help determine how much of an ore resource is classified as a reserve.) By contrast a resource is a known deposit that, at current market prices, is not economically recoverable, i.e., profitable to the producers. Of a still more hypothetical nature are "Potential Resources." These are quantities of uranium which are thought to be available, although we do not necessarily know either the location or size of the conjectured resource. Sometimes these "Potential Resources" are further subdivided into "Probable," "Possible," and "Speculative" categories, indicating steadily increasing degrees of uncertainty with respect to the economical recovery of the mineral and even to the very existence of the resource.

ERDA estimates that in addition to the 690,000 tons of uranium reserves, the U. S. has potential resources equal to less than three million additional tons of postulated uranium.[17] For both its reserve and resource categories, ERDA has regarded as economically recoverable only ores whose final production cost can be held below $30 per pound. Most of these potential resources, however, can only be obtained from very low grade uranium and it is wistfully optimistic to expect that more than a small fraction of these potential resources will ever be recoverable for prices anywhere near $30 a pound. The distinction between cost and price here is an important one: cost means the expense of production whereas price means cost plus an allowance for industry profit.

Unfortunately for the nuclear industry, the 690,000 tons of known uranium reserves that the U.S. is counting on are sufficient to fuel no more than sixty-two large nuclear reactors (1000 MW each) throughout their assumed maximum forty-year operating lifetimes!*

The detailed assumptions and calculations on which this conclusion rests may be found in Appendix A: Uranium Fuel Duty. The result was derived by Mr. Morgan G. Huntington, an engineer employed by the U.S. Mining Enforcement and Safety Administration of the U.S. Department of the Interior. This writer is greatly indebted to Mr. Huntington for his analysis and interpretation of uranium fuel supplies, although responsibility for this account of his work is the author's.

Because only sixty-two plants can be fueled with certainty, a great deal more uranium ore will be needed for the U.S. reactor program than ERDA and the nuclear industry are willing to admit.

Prospects for fueling more reactors would be much better if current U.S. uranium reserves were rapidly increasing to meet the growing demand. But instead, strong inflationary effects are removing large amounts of uranium from various reserve and resource categories. The U.S. from 1974 to 1975, for example, actually lost 77,000 tons of uranium from ores in the $8 recovery cost range, and similar but lower losses were sustained in the higher price ranges.[18] Actual additions to reserves during the same period, 13,000 tons, were just balanced by the 12,600 tons of uranium mined and shipped to mills. Thus while new reactors went on line and other approached commercial readiness, U.S. uranium reserves actually shrank.[19]

ERDA's reluctance to admit how large the uranium shortfall will be is understandable, in view of the fact that recovery of the necessary amounts of uranium from low-grade ore will yield *trivial* amounts of net energy and will be *economically unfeasible,* costing a *minimum* of a hundred dollars per pound. That cost is more than *three times* the maximum amount which the federal government has used as a cut-off point in computing uranium reserves.

---

*No commercial reactor has been around long enough for us to know how realistic such estimates may be.

This simple result — coupled with the other high costs of nuclear power — has *devastating* consequences for the nuclear industry. Those consequences are outlined below, followed in a moment by an account of their derivation.

1. *Commercial nuclear power in the U. S. is not an economically viable option once high-grade uranium is exhausted.*

2. Depletion of high-grade uranium will occur *long* before all the reactors projected by ERDA can be fueled.

Although the U. S. has assured supplies of high-grade uranium for only sixty-two reactors throughout their lifetimes, ERDA has forecast that 725 reactors would be in operation by the year 2000. The AEC, before its demise, had projected a thousand. It follows from the preceding discussion and the calculations of Figure 1 that the U. S. will need *more than eight million tons* of uranium to fuel 725 reactors during their forty-year lifetimes. Not only is that quantity many times larger than current reserves — it is more than twice the size of current uranium reserves plus all the probable, possible, and speculative uranium *resources* combined and with all the byproduct uranium from copper and phosphate mining from now to 2020 thrown in for good measure!

Moreover, even if the U.S. met with colossal success in uranium exploration and thereby managed to find uranium reserves totaling a million or two million tons, at least six or seven million tons of uranium still would have to be obtained from low-grade ores whose recovery is not only uneconomic, but totally unfeasible. To get those six or seven million tons, six thousand to seven thousand square miles of land would have to be mined — an area well in excess of that mined for all mineral commodities in the U.S. today. And the net energy yield of such an operation would be paltry at best. Therefore, it must be the dollars they are after. Apparently the nuclear industry is quite willing to bury us in mine tailings in their quest for profits.

3. The economy simply cannot derive a large portion of its energy from nuclear fission by the year 2000, as its proponents claim. *All known uranium reserves, if fissioned in light water reactors, can never on the average produce more than a gross 5 percent of U.S. annual energy requirements during the reactors' forty-*

year predicted lifetimes.* If from the *gross* energy yield we subtract all the energy inputs needed (to mine the uranium ore, build the plant, etc.) the *net* contribution of nuclear power is even lower.

Nuclear advocates have asserted that without nuclear power, the economy will run short of energy and that the scarcity will create unemployment. But the facts above clearly imply that nuclear energy cannot by any stretch of the imagination be considered as a form of job protection. Moreover, the little energy we do get from nuclear power would actually produce far fewer jobs than many energy alternatives (see Part II, this book).

It is a *physical impossibility* for the U. S. to obtain a great deal more energy than this from nuclear fission barring a series of high-grade uranium strikes of gigantic size, each equal to the multistate Colorado plateau region. Even one such find would in itself be so improbable as to defy description.

4. Far from being essential to our economy, building more and more nuclear reactors *on the hypothesis* that enough high-grade uranium will be found in the future to fuel them is the epitome of economic irresponsibility.

If several hundred reactors were built as projected, the nuclear industry would find itself dependent on the low-grade uranium ores found in the widely dispersed shales of the Devonian period (commonly known as Tennessee shales). This would cause tremendous uranium price escalation,† resulting in large price increases for the overall system cost of nuclear power. As a result, consumers' electric rates would skyrocket, and the utility industry would suffer, with disruptive economic consequences for the rest of the economy (especially among those economic interests with financial commitments to nuclear utilities and reactor builders). Further nuclear expansion will, therefore, worsen the lot of those

---

*Total U.S. reserves of 690,000 tons of uranium when fissioned in current reactors can yield no more than 3.91 quads of energy every year for the forty year lifetimes of those reactors. With current U.S. energy use at about 75 quads in 1973 and growing, $\frac{3.91}{75}$ yields 5.2 percent. (A quad equals 10^{15} British thermal units) For further explanation, consult Technical Appendix A.

†Undoubtedly just to the point below which it would be cheaper for the utilities who had already bought nuclear plants to buy the costly fuel rather than to close down the plants entirely and write them off as billion dollar capital losses.

workers who are already poor today and not protected by lifeline utility rates. The beneficiaries of this situation will be those large energy firms which have cornered much of the country's scarce uranium reserves.

5. Neither the introduction of nuclear fuel reprocessing nor the breeder reactor by 2000 A.D. will substantially alter these consequences, and neither of these technologies is likely to be widely introduced. In fact, the implications of the fuel shortage are likely to force curtailments in the breeder program itself.

DERIVATION OF THE HUNTINGTON RESULTS

Morgan G. Huntington's major conclusions on the potential energy available from uranium and the fuel-supply implications of those conclusions follow as a direct consequence of his calculation in Chart 7-1 (page 160) that one ton of uranium oxide (U_3O_8) can *at best* produce a gross electrical energy yield of no more than 22 million kilowatt hours (electric) when fissioned.[20] This implies that each 1000 Mw reactor would require 11,150 tons of uranium oxide during the course of its forty-year lifetime, if it were to deliver its promised quota of electricity. Because one ton of uranium oxide will make .12 ton of 3-percent enriched reactor fuel, 11,150 tons of uranium oxide will be sufficient to produce 1,305 tons of reactor fuel — enough to fuel one reactor for forty years on a normal fueling schedule.

It is also apparent that because U.S. uranium reserves contain no more than 690,000 tons of uranium and each 1000 MW reactor requires 11,150 tons, no more than sixty-two large plants are fuelable throughout their forty-year life expectancy. (Division of 690,000 tons by 11,150 tons equals 61.9.)

The nuclear industry has repeatedly used far higher estimates of the gross energy yields possible from a ton of uranium. ERDA has used the figures 32 million kWh/ton and even 70 million kWh/ton.* Dr. Hans A. Bethe, one of the early advocates of commercial nuclear power (and a champion of it today) has recently claimed 70 million kWh/ton of uranium (*Mining Engineer,*

---

*Neither figure assumes fuel reprocessing, which could only add 36 percent to the gross energy yield.[21]

August 1975); nuclear advocate Dr. David J. Rose, Professor of Nuclear Engineering at MIT, publicly claimed 51 million kWh/ton.

But in actual performance, contemporary reactors have not even come up to Huntington's gross energy estimate of 22 million kWh/ton, much less to 70 million kWh. Operating experience as tabulated by Huntington for U.S. reactors from 1970-1975 reveals that observed fuel duties were in the range of only *6 million kWh* of electricity per ton of fuel placed in the reactor!

Moreover, both Huntington's theoretical calculations and the much lower actual operating experience are estimates of the *gross* energy production of uranium fuel in reactors. Considerable energy has to be expended in mining, milling, processing, transporting, and fabricating the uranium fuel. After subtracting the energy inputs absorbed by these activities, Huntington found each short ton of uranium fuel could yield a maximum of only 18.5 million kWh in net energy. (About 3.5 million kWh were required as energy inputs per ton).

Because in any energy production technology the quantity of paramount importance is the *net* rather than the gross energy produced, it is apparent that out of 6 million kWh of gross energy per ton produced by U.S. reactors so far, only 2.5 million kWh per ton were net energy gained!

By the methods shown on page 161, it is evident that if actual nuclear plants only produced 6 million kilowatt hours of gross electrical yield per ton of uranium, that would amount not to an average 5 percent of annual energy over forty years, but to only 1.4 percent gross energy contribution per year, and to a *net* energy contribution of less than 1 percent from uranium fission.

This pitifully low energy yield makes uranium fission unfeasible as a major energy source and a most expensive form of limited energy. It also explains why the U.S. government and the nuclear industry have been extremely reluctant to furnish complete energy yield data on operating nuclear power plants. To call uranium fission a practical or an economical source of energy is clearly wrong. And to build more light water reactors is both economically foolish and futile — it will only cause ever higher utility rates.

By assuming greater energy yields from uranium fuel than are even theoretically possible today, the nuclear proponents have concluded that less uranium is needed to fuel the reactors they plan to build than will really be needed. For example, if uranium fuel duty (energy yield) is 44 million kWh instead of the 22 million kWh derived by Huntington, then each reactor needs only *half* as much fuel as it would if it were generating only 22 million kWh. Similarly, if uranium fuel duty is 70 million kWh, then only a *third* as much fuel is needed.

Even assuming that commercial fuel reprocessing were implemented — and it has formidable political opposition and technical difficulties — this would add at most about 36 percent to the gross uranium fuel duty, according to Huntington. Instead of fueling sixty-two reactors, at most about eighty-four could be fueled. Of course, this is contingent on both Nuclear Regulatory Commission approval of reprocessing and the successful operation of a commercial U.S. reprocessing facility. So far, the technology has not been proven to be more than "marginally economic."* The value of uranium-235 recovered is greatly lowered by the inevitable presence of uranium-236 in the spent fuel; similarly, the recovered plutonium-239 is contaminated by plutonium-240.[22]

In learning the true situation concerning the U.S. uranium scarcity, people naturally wonder whether the uranium shortage could not be alleviated by simply searching harder for domestic high-grade uranium.

The answer is that although we are increasing our rate of exploration, the outlook for large new strikes of high-grade ore is poor. Rich sources of uranium are becoming much harder to find, and mineral returns on exploratory activity have already greatly diminished. Test bores drilled in the 1950s, for example, averaged about one hundred sixty feet in depth before uranium was found. Just a decade later, drillers were having to bore more than four hundred feet on the average just to hit ore, which in general was too poor in quality to make its recovery worthwhile. (Recall the definition of a resource.)[23]

Using the resource depletion estimation method of former Shell Oil Co. geologist Dr. M. King Hubbert, Huntington applied those techniques to uranium exploration records showing the

*David Burnham, "Hope for Cheap Power from Atom is Fading," *New York Times*, November 16, 1975.

quantities of uranium found per foot of exploratory bores drilled. The data clearly indicated that the average discovery rate per foot of drilling activity for high-grade uranium ore had fallen by more than a factor of ten since 1948. Projecting from the drilling data, Huntington estimated that total U.S. uranium reserves will not exceed 744,000 tons of high grade ore for all time.[24] And of these 744,000 tons, 600,000 tons have already been found in ores richer than 200 parts per million (ppm), of uranium with only 144,000 tons remaining to be discovered in ores of this grade. To nullify Huntington's fuel projections, uranium explorers in the U.S. would have to make several enormous finds of uranium between now and the year 2000. Yet ERDA uranium expert Robert Nininger has stated, "No new major uranium producing areas or potential producing areas have been identified in this country during the past seventeen years."

Nevertheless, ERDA is reluctant to relinquish the will-o-the-wisp of cheap nuclear energy and refuses to acknowledge that uranium fission can offer the nation only trivial quantities of expensive energy. In an effort to salvage the disastrous situation into which the AEC and ERDA have plunged the American consumer and the nuclear utilities (most of whom currently only hold contracts for a few fuel loads), ERDA has mounted a desperate multi-million dollar search for uranium throughout every corner of the U.S.

Known as the National Uranium Resource Evaluation (NURE) Program, ERDA will try to augment information on uranium deposits by means of an aerial "radiometric reconnaissance program" and a national "hydrogeo chemical" survey. Eight million dollars was budgeted for the program in 1974 and 1975 combined, and $14 million is budgeted for 1976. Yet no matter how much money ERDA pours into such high technology surveys and no matter how many computer printouts are generated, it cannot alter the basic fact that high-grade deposits of uranium ore close to the surface of the earth are relatively rare accidents of geology.

In addition, ERDA's estimates of known uranium reserves create the impression that our reserves are more adequate than they really are. A study of U.S. uranium reserves producible at up to $25 a pound, by noted Canadian geologist Dr. Frederick Q. Barnes, revealed U.S. reserves total only *308,000 tons* (*Mining*

Congress Journal, February 1976). And some of these reserves will require decades for their exploitation; they will not be accessible *until well within the next century.* As Dr. Barnes has pointed out, potential production capability is an important limiting factor that reduces the immediate utility of reserves. The annual required discovery rate for the U.S. in 1975 was 60,000 tons of U_3O_8, Dr. Barnes concluded, while potential production capability was only 13,200.

Given the shortage of high-grade ore and the likelihood that it will continue, it is natural to wonder whether the U.S. could not just import enough high-grade uranium from abroad, or use medium- or low-grade domestic uranium deposits? Much uranium could have been imported in the 1960s, but in a mood of ill-considered optimism about domestic uranium reserves, the U.S. banned uranium imports in 1964 to protect the domestic uranium industry. That embargo on imports is being allowed to expire in 1977 because government planners have finally realized how scarce our domestic uranium supplies are. But their realization has come too late. According to ERDA, "Over the long term, prospects for significantly augmenting U.S. uranium resources (sic) with imported uranium are not good, unless new discoveries add appreciably to currently estimated foreign resources. The installation of nonbreeder nuclear capacity abroad is proceeding at a rapid pace and is projected to grow such that all currently estimated foreign uranium resources (noncommunist countries) evidently will be needed to support this nuclear capacity."[25] Translation: the nuclear technology which the U.S. was so instrumental in spreading around the globe has now begun to compete with U.S. utilities for scarce world supplies of uranium. But even if large new supplies of foreign uranium were discovered as a result of intensified exploration, how could the large-scale importation of uranium be reconciled with earlier ballyhooing for nuclear power as the high road to "Energy Independence"? Could it be that dependence on a uranium cartel or foreign uranium exporters for expensive uranium would in some mysterious way be preferable to depending on the Organization of Petroleum Exporting Countries? Not unless the "high road" is paved with bricks of Fool's Gold.

As for the suggestion that the uranium scarcity could be alleviated by relying on medium-grade ores, the problem is simply that relatively little intermediate-grade ore exists.* Once high-grade fuel is committed to existing reactors, the U.S. is then consigned to get any additional uranium from the low-grade ores, which the U.S. possesses in great abundance, just as there are even vaster amounts of highly dilute uranium in seawater. The catch is money.

The concentration of uranium in low-grade ores is greatest in shale deposits, but the ore is so diffuse it exists only in concentrations of sixty to eighty parts per million. In other words, for every ton of rock mined, only 2.24 ounces of uranium are found.

If it were economically possible to recover this uranium oxide, one would have to mine four tons of rock just to eventually get the same amount of energy from fissioning the uranium as could be obtained far more easily by mining only one ton of coal!

Why would anyone want to go to the trouble and great expense of mining four tons of rock and then subjecting it to so much complex processing to get uranium (worth less than 836 kwh net), when greater energy value could be had by mining far less coal? Not only does the coal, pound for pound, have about four times larger net energy value than the low-grade uranium ore, but in place of the long, involved uranium processing, all the coal needs is crushing and washing.

Once the uranium oxide is mined from the ground, a great many things remain to be done before it can be fissioned and even after the ore is refined, only a tiny fraction can actually be fissioned.

A ton of low grade uranium ore with 70 parts per million uranium contains only 2.24 ounces of uranium. About 30 percent of that 2.24 ounces is discarded with mine tailings at the mine site leaving 1.57 ounces of uranium. That metal, however, contains only 0.7 percent fissionable uranium-235 (the rest is unfissionalbe uranium-238). Although we now have .011 ounces of fissionable material, roughly another third of that is lost in "enrichment tails" during enrichment of the uranium to 3 percent ura-

*ERDA acknowledges the absence of intermediate grade ore:

"There is an apparent gap between sandstone ores and Chattanooga shales in terms of large resources of intermediate uranium content. Future exploration may uncover material in the intermediate range, but for the present the gap is real enouch."[26]

nium-235. This leaves only 0.0074 ounce of uranium-235, of which only 85.5 percent will fission in a reactor because some of it will be converted to uranium-236. Of the remaining uranium-235 that will fission, only 70 percent is normally "burned up" before a reactor is refueled. Therefore, out of the original ton of rock mined, only a grand total of forty-four ten thousandths (0.0044) of an ounce of uranium-235 can actually fission to produce electricity, supplemented by an "energy bonus" of about 43 percent from the fissioning of other isotopes. As one would expect and as earlier calculations have shown (see page 160), the net energy so derived from such low-grade uranium is *extremely low* (only 836 kWh). And the cost of getting that energy would be extremely high, if anyone were to bother doing it.

Dr. John Klemenic, Acting Director of ERDA's Supply Analysis Division in Grand Junction, Colorado, computed the cost of recovering uranium from ore bearing 100 parts per million (ppm) of uranium *oxide* at $100 per pound, assuming open pit mines that produce five thousand tons a day.[27]* Making an allowance over production cost for a 12 percent cash flow rate of return on producers' investments, Dr. Klemenic found that the sale price for that pound of uranium oxide would be $120. However, the ores of the Tennessee shale formation's richest member, the Gassaway formation, contain only 70 ppm of uranium oxide (59 ppm uranium).[28] When Dr. Klemenic's cost estimates are applied to that shale, ore costs leap to $143 per pound. Assuming that the same return to investors produces roughly a one fifth increase in price over cost, the sale price for uranium oxide increases to $172 a pound.

If uranium developers were able to exact, say, a cash flow rate of return of 27 percent from fuel hungry utilities then, by the methods above, the sale price of that uranium would jump to about $229 a pound. And when these cost projections (of $172 and $229) are modified to reflect two years of inflation from 1974, when Dr. Klemenic wrote, to the present, we discover that low-grade uranium-bearing shales from open pit mines are liable to produce uranium for $193 or even as much as $257 a pound, depending

*Of the 100 ppm U_3O_8, only 84.8 will be uranium metal.

on the profit rate.* These results assume a strict proportionality between uranium ore grade and recovery costs per pound of uranium. Actually, in the recovery of low-grade ores, costs increase *more* than proportionally, because ore recovery processes become less efficient as ore grade diminishes.

Those readers familiar with the nuclear fuel cycle may wonder why we have considered resorting to low-grade shales for uranium when the U.S. has available mountains of discarded uranium tailings produced in milling the rich uranium ores already used to build bombs and to fuel reactors. These tailings themselves contain three times as much uranium *per ton* as do the shales we have been discussing. But the mill tailings do not exist in quantities anywhere near large enough to matter greatly in producing uranium-235 reactor fuel. Tailings from the approximately two hundred thousand tons of uranium oxide that have been milled to date, with a 95 percent recovery rate, contain only about ten thousand tons of uranium oxide.

Even though fuel has traditionally been a small part of the overall cost of delivered nuclear power, fuel price increases of the magnitude just discussed would very definitely make uranium fission economically uncompetitive with coal and many other energy sources. As explained, a market is unlikely ever to develop for such expensive uranium. Therefore, ERDA's plans for hundreds of reactors fueled by uranium 235 fission are simply pipe dreams.

Table 7-3.

ERDA Estimates of Recovery Costs for Low-Grade Uranium Resources

Host deposit	Tons of U_3O_8 (millions)	Grade ppm	Forward cost
Shale	5	60 to 80	$100
Shale	8	25 to 60	150
Granite	8	10 to 20	200
Shale	200	10 to 25	200+
Granite	1,800	4 to 10	200+
Seawater	4,000	0.003	500+

Ore grade in parts of U_3O_8 per million of host. Cost in dollars per pound.

---

*Dr. Klemenic found that low-grade uranium procured from underground mines would be only about 16 percent cheaper than from open pit mines for ores of 100 ppm uranium oxide.

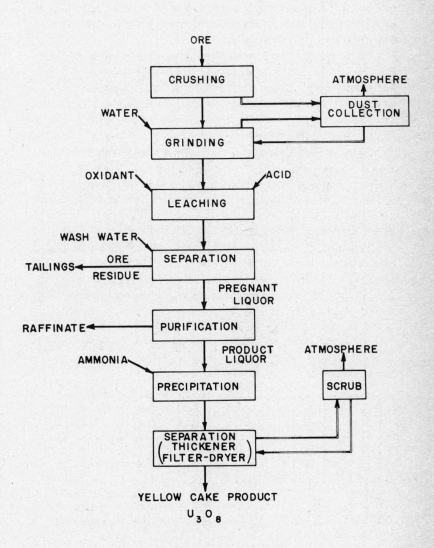

Figure 7-1. Uranium milling plant operations

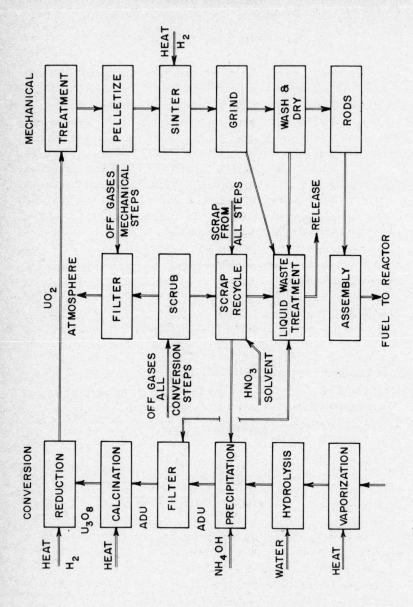

Figure 7-2. Fuel fabrication ADU process, simplified flow diagram

THE BREEDER REACTOR "SOLUTION"

The breeder reactor has often been presented as the ultimate answer to the uranium shortage. For example, Dr. Chauncey Starr, president of the Electric Power Research Institute, once boldly claimed, "The breeder reactor should make it possible for nuclear fission to supply the world's energy needs for the next millennium."[29] In theory, the breeder is a wonderful way to produce an almost infinite supply of nuclear fuel. When fueled with an initial charge of up to four tons of toxic plutonium, the breeder can convert plentiful uranium-238 to plutonium-239, which itself can be fissioned to produce electricity. Although the nuclear establishment is straining to develop a working breeder, breeder prospects are not comforting and offer no fuel supply panacea.

First, ERDA does not plan to have a breeder in commercial operation until sometime in the 1990s, and the U.S. is unlikely to have more than a few token breeders before 2000, assuming all goes well for the current U.S. breeder experiments. Next, even a commercial breeder in the early 1990s, however, "would have no significant effect on uranium requirements in the year 2000,"[30] according to Dr. Ralph Lapp, a prominent nuclear proponent.

Referring to the breeder's "self-fueling nature," Dr. Lapp appears to strongly support early commercial breeder introduction. Yet the French Phenix breeder reactor, the only breeder in the world now operating at full power and the only one for which doubling time* data are available, will take forty to sixty years before it will double its fuel load.[31] (See page 91.) With the help of many technological improvements, the French may be able to reduce this doubling time to thirty to forty years. Thus, if a generation of Super Phenix breeder reactors were fueled and started up in the year 2000, they might double their fuel charges of plutonium by 2030. That would be a little late to fuel reactors coming on line this year or in 1985. Moreover, breeder reactors are currently an *unproven technology* and a great many economic and safety objections have been raised about them. It is by no means certain that the U.S. will ever have a large number of breeder reactors. There-

---

*The time needed for the breeder to produce enough fissile material to double the reactor's total stock of fissionable fuel.

fore, to say the least, we cannot count on this future technology to
solve the current fuel shortage.

In addition, it follows from Huntington's data on uranium
fuel duty that, given the scarcity of high-grade uranium and the
economic barriers to using low-grade ore, the nation will not be
able to fuel enough light water reactors to produce even the amounts
of plutonium needed by the breeder-reactor program envisaged by
ERDA. With uranium reserves of 690,000 tons and sixty-two fully
fueled reactors, about six hundred twenty tons of plutonium will
be produced during the reactors' forty-year lifetimes.* Assuming
each breeder takes a four-ton core load of plutonium only one
hundred fifty-five breeders could be fueled, as opposed to the five
hundred forecast by ERDA. And, of course, even the lower projection assumes that the breeder's difficult economic, safety, and
environmental problems be successfully resolved, and that fuel reprocessing proves environmentally safe and legally acceptable.

THE ENRICHMENT BOTTLENECK

Escalation in the costs of bringing uranium out of the ground
is only part of the utilities' increasing fuel costs. Once uranium
ore is mined and milled into a fine yellowish sand (yellowcake),
the uranium must be converted to uranium hexafluoride (UF_6) so
that it can be enriched to a fissionable grade. And those enrichment costs are also rising. Possible increases in enrichment costs
have not so far been included in our estimates of uranium fuel
price escalation.

To date, the U.S. government has a monopoly on U.S. enrichment and provides enriched uranium for nuclear fuel from
three uranium enrichment plants at Portsmouth, Ohio; Paducah,
Kentucky; and Oak Ridge, Tennessee. Originally built to enrich
uranium for atomic and hydrogen bombs, these plants do not have
enough capacity both to support a full-scale domestic nuclear industry and also to provide enough enriched uranium for U.S. reactor sellers to use in sweetening their foreign sales contracts.

The plants have already committed their entire enrichment

---

* According to conventional estimates, each 1000 MW reactor at currently observed
operating capacity produces about five hundred pounds of plutonium per year.

capacity and ERDA is planning to expand them at a cost of a billion dollars. But even the projected 60 percent expansion is already fully committed to ERDA customers in the U.S. and abroad,[32] so ERDA is strongly supporting President Ford's controversial efforts to get private enterprise into the enrichment business, a step that would very likely raise enrichment prices considerably.[33] These prices are already on the upswing.

In 1975, ERDA raised enrichment prices from $36/SWU* set by the AEC in early 1974 to the $60 range, and increases to $76 are in view. If the gap in enrichment capacity is met by a commercial enrichment facility as has been proposed, the price for enrichment could rise to $100/SWU or more. A large diffusion enrichment plant would cost $3.5 billion and it would draw colossal amounts of electric power. Several new enrichment plants might be needed to meet the nuclear power industry's uranium fuel needs.

ERDA Deputy Administrator Robert Fri has estimated that to meet current domestic and foreign enrichment requirements, the U.S. should spend $30 billion by building up to twelve new enrichment plants. *Wall Street Journal* reporter Jonathan Kwitny states that the cost of expanding nuclear fuel production will be closer to $60 billion.[34]

Massive government subsidies from ERDA have been suggested to insure that commercial enrichment facilities are built, but there is no guarantee that this will happen at all or that, if it occurs, sufficient enrichment capacity will be available in time.

Leading contender for the task of expanding U.S. enrichment capacity is a consortium of three firms known as Uranium Enrichment Associates (UEA). Composed of Bechtel Corporation, Goodyear Tire and Rubber Co., and Williams Co. (a large pipeline company that also manufactures pumping equipment), the consortium is proposing to build a $3.5 billion gaseous enrichment plant near Dothan, Alabama, and to construct two new nuclear power plants costing another $2.2 billion to provide power for the enrichment facility. President Ford has extended strong support for the $5.7 billion project and proposed that an $8 billion fund be created to rescue companies like UEA in case they encounter financial problems in their enrichment ventures.

---

*SWU stands for "separative work unit," which is the amount of energy needed to enrich a kilogram (2.2 pounds) of natural uranium to 3 percent U-235.

Bechtel Corporation, a founder of UEA and one of the world's largest engineering and construction companies, has hired high former officials of both the AEC and the Nixon administration as it has proceeded with its enrichment plans. Bechtel officials deny that there is any connection between the proposed enrichment project and the hiring in 1974 of Robert Hollingsworth, former general manager of the AEC; George Schultz, former Secretary of the Treasury in the Nixon administration; and former Secretary of Health, Education, and Welfare, Caspar Weinberger, who was hired in 1976 as a vice president and special counsel. Mr. Schultz was also hired as a vice president, but he has now become president of Bechtel.

REPROCESSING

In addition to uranium ore and enrichment costs, further costs will occur after uranium is fissioned in a reactor. During the fissioning, fissionable plutonium-239 is created and some uranium-235 remains unfissioned. Both can be recovered during reprocessing, provided one is willing to pay a high dollar cost for a small energy return. Ordinary reactors could fission mixtures of plutonium and uranium fuel, but this would increase the hazards of the nuclear fuel cycle (as explained in Chapter 5).

At one time, many utilities expected that plutonium recycling would be instituted virtually automatically and would generate additional earnings to defray earlier fuel cycle costs. However, Bertram Wolfe, General Manager of Fuel Recovery for General Electric and a coauthor, Ray W. Lambert, have concluded that nuclear fuel reprocessing may not be commercially viable:

> In the past, the back end of the fuel cycle was looked upon as a profitable part of the fuel cycle; that is, one expected that the value of the plutonium and the uranium would be significantly higher than the recovery, fabrication, transportation and waste management costs. *This is not necessarily true any more.* [35] (emphasis added)

This finding was independently confirmed by physicist Marvin Resnikoff of the State University of New York at Buffalo.[36] Resnikoff's results depend on the fact that recovered uranium-235 is contaminated by uranium-236, which absorbs neutrons in the reactor and slows the fission process, making the recovered fuel much less valuable than an equivalently enriched batch of fresh fuel. The AEC, however, for years encouraged the construction of reprocessing facilities by giving nuclear utilities financial credit on their enrichment bills for uranium that one day might be recovered from the utilities' irradiated fuel. The provision of these credits was based on the assumptions that (1) plutonium recycling would be approved, (2) uranium would be recovered and become AEC property, and (3) that the contaminated fuel was *equal in value* to the equivalent amount of fresh uranium.[37]

Nuclear industry planners have estimated that just to service two hundred nuclear plants, four large reprocessing plants may be necessary by 1990 at a cost of $4 billion. Meanwhile, the country's newest, largest, and still unopened Barnwell reprocessing plant has been so fraught with delays, cost overruns, and regulatory uncertainties that Allied-General Corporation has asked for federal help. Even if Barnwell begins operating on schedule and the Nuclear Fuel Services plant in West Valley, N.Y., reopens, spent-fuel production will have exceeded the capacity of both plants by 1984.*[38] Further bottlenecks are therefore almost guaranteed in waste-reprocessing and management: It takes about a decade to construct a major facility, and, with Allied-General's troubles still vivid in the minds of nuclear experts, it will probably be a long time before another firm enters the business. That leaves the federal government, in other words, the taxpayer, holding the nuclear trash barrel.

NUCLEAR PLANT PERFORMANCE

An extremely important determinant of overall nuclear electric costs is a nuclear plant's capacity factor. Simply put, capacity factor is a measure of the average amount of power that is produced by a reactor (or any power plant), expressed as a percentage of the maximum amount of power that plant *could* have produced under optimum conditions:

$$\text{Capacity factor} = \frac{\text{Actual power production}}{\substack{\text{Maximum total power of} \\ \text{which plant is capable}}}$$

The top half of this fraction is given by how many kilowatt hours a plant produces. The bottom half is generally given by the specifications to which the plant was originally built, known as its design-rated capacity, (e.g., 1000 MW), multiplied by the number of hours the plant could have operated if it had been performing perfectly.

Plant capacity factors can be lowered in only two general ways. Either the plant is unavailable to produce power for a portion of its operating period, because of mechanical problems, or the plant may operate for the entire year, but may be only generating a fraction of its design-rated power. For example, if a plant with a 1000 MW rating operates 100 percent of the time at 60 percent of that capacity, its average output is 600 MW. Should it operate for nine months at 60 percent of capacity and be shut down completely for three months for mechanical repairs, the plant's capacity would be 45 percent — by making a weighted average of nine months capacity at 60 percent and three months at zero capacity.(¾ year x 60 percent + ¼ year x 0 percent = 45 percent)

Because reactors sometimes operate at less than maximum power and because they shut down periodically for repairs and refueling, it is nearly impossible to operate them for long at 100 percent capacity. When the AEC made its final environmental impact statement calculations about the probable power output and costs of all U.S. nuclear power plants, the Commission assumed the plants would be able to operate at 80 percent of capacity. Only belatedly, shortly before the AEC was dissolved, did it modify its expectation to 75 percent. But in the very first year when the AEC published capacity factor data for nuclear power plants (May 1974), it soon became clear that actual capacity factors were far below the 75 to 80 percent range.[40]

Those AEC capacity data were analyzed by David Dinsmore Comey of the Chicago-based Business and Professional People in the Public Interest. Comey found that typical U.S. reactors actually operated at only 57.3 percent of their capacity in 1973.[41] From 1973 to 1975, Comey found that nuclear plants in the U.S. averaged only 55.1 percent of their design capacities. Two important facts emerge from Comey's data:

First, the average capacity figures are substantially worse than the AEC assumed (for plants of all sizes), and plants larger than

1000 Mw in particular show even poorer-than-average capacity factors of 44.5 percent — more than 20 percent lower than plants in the 400-499 MWrange.

Such results are especially disturbing to the nuclear industry in that most new nuclear plants being built are large ones. If the results extend generally from the six plants in Comey's survey larger than 1000 MW to all future large plants, the nuclear industry is in terrible straits indeed. Those results would imply that to get 1000 MW of power, it would be necessary to build two 1000 MW plants, not one. Naturally that would double the capital cost of nuclear electricity. Such a conclusion is still premature because of the small number of plants larger than 1000 MW studied by Comey.

Secondly, Comey's figures also suggest that nuclear plants' performance begins declining far earlier than the AEC forecast. In *Nuclear Power Growth 1974-2000*, the AEC projected that nuclear plants would perform at an average of 65 percent capacity during the first thirty years of operation;* furthermore, the AEC projected that performance would drop from an expected 75 percent capacity during the fourth to the fifteenth operating year to about 39 percent capacity *only by the thirty-third year of plant operation.* [42] But the Comey data show that for the three oldest commercial plants that operated for all of 1975, average capacity factors had fallen to 39.2 percent after only twelve to fifteen years of service — in less than half the time the AEC had projected.

If this trend occurs as a general rule for other nuclear plants, too, as they age, then nuclear plants will never on the average attain an overall annual capacity figure of 65 percent during a thirty-year lifespan, as the AEC stated. Two Swedish nuclear engineers, Peter Margen and Soren Lindhe have proposed an alternate model of plant performance to the AEC version. Although the engineers are strong nuclear proponents, their results still imply that nuclear plants will average only about 43 percent capacity during a thirty year plant life.[43]

POWER PLANT AGING

In 1975 seventy-three out of seventy-five new nuclear plants under

---

*The AEC assumed nuclear plants can operate for a total of forty years.

construction in the U.S. were larger than 800 MW. Because the ten largest U.S. reactors had not been in service more than three years in 1974, their performance could improve considerably later. But even if the plants did perform at improved capacities for their few years of operation (as they were "debugged"), there still would be no reason to assume that the plants would be immune to the capacity-draining effects Comey found for all plants. As Comey explains, "Since the radioactivity of these plant systems increases with plant age, repairs are likely to become even more time-consuming as the plant gets older, leading to longer outages and decreased capacity factors."[44] Because each worker can be exposed to no more than five rems of radioactivity in a year, except under specified circumstances when five to twelve rems per year might be allowed, repairs on highly radioactive parts of the reactor can take thousands of men and many times as long as a comparable repair on a fossil-fuel plant.

A simple example will help translate some of this capacity factor information into dollars and cents. If utilities spend $100 billion to build a hundred 1000 MW nuclear plants on the assumption that they will function at 65 percent capacity and generate an average of 65 GW(e) (65 billion watts), the utilities are buying power at the rate of $1.54 billion per gigawatt. But if for that $100 billion, those hundred plants deliver only 43 percent capacity, or 43 billion watts, the utilities have had to spend $2.33 billion per gigawatt. In order for the utilities to buy 65 GW at that rate, they would have to spend not $100 billion but $151.5 billion on nuclear plants. Considered another way, utilities who paid $100 billion thinking they were going to get 65 GWe only to get 43 GWe have overpaid the nuclear reactor vendors by an extra $33.78 billion.*

Another important analysis of nuclear capacity-factor data has been performed by the Council on Economic Priorities, a public-interest research group based in New York City. Headed by the council's Energy Policy Director, Charles Komanoff, the study's results are consistent with the Comey data on the negative correlation between increasing plant size and plant capacity. CEP found statistically significant evidence that for each 100 MW increment in nuclear plant size, plant capacity factor declines by 1.5 percent.

*65 GW x $1.53 billion = $100 billion; 43 GW x $1.54 billion = 66.22 billion; $100 billion —$66.22 billion = $33.78 billion.

The study also confirmed that capacity factor declines at about 1 percent for each yearly increase in plant age.[45]

THE NUCLEAR "COST ADVANTAGE"

Although the information already presented shows how power from uranium fission will become far more expensive than coal power, until recently it was difficult for many economists to predict whether coal or nuclear power would be a cheaper source of electricity in the long run. But by the mid-seventies, information on nuclear power's escalating costs had become increasingly evident. Simultaneously, the nation's powerful pronuclear lobby mounted vigorous efforts to mislead the public about this matter. Consolidated Edison Co., for example, which was obliged in 1975 to sell its third nuclear power plant to the State of New York because of the company's economic difficulties, valiantly tried to justify its nuclear commitment to the public.

Another example, in January 1975, Con Edison announced in a news release, "[Our] customers benefited from nearly $100 million in lower fuel charges in 1974 because of the operation of [our two] nuclear facilities at Indian Point at Buchanan, N.Y."[46] In its "Customer News" pamphlet mailed with April 1975 electric bills,[47] the company reiterated that "operation of Con Edison's nuclear plants in 1974 saved our customers $95 million they would have otherwise paid for an equivalent amount of oil." The intent of the announcements was clearly to create the impression that nuclear power plants, through their "fuel savings," were actually on balance saving customers money.

However, a detailed study of the Indian Point nuclear plants by the Council of Economic Priorities concluded "the Company's commitment to nuclear power at Indian Point 1 and 2 caused little, if any, reduction in the cost of electricity in 1974 and may even have increased the cost of electricity to its customers."[48] Charles Komanoff, the study's author, concluded that had the company built a coal-fired plant instead of the Indian Point 2 plant, the coal plant would have been $26.8 million *less* expensive to own and operate in 1974 than Indian Point. Komanoff also computed the cost of the coal-fired plant with scrubbers added,* and found

---

*Scrubbers are useful on coal-fired plants to reduce emissions of harmful sulfur oxides through the plant's stacks.

the plant would still have been cheaper to own and operate in 1974 than the nuclear plant (by at least $17.9 million). The Council on Economic Priorities report was submitted to Representative Richard L. Ottinger (Dem.-N.Y.) who sent it to the Federal Power Commission, the Federal Trade Commission, and the New York State Public Service Commission, asking them to investigate Con Edison's nuclear advertising. As of late 1975, no action had been announced.

Another example showing how the advocates of nuclear power are using highly unrealistic cost estimates to make nuclear power look competitive is the case of the Rancho Seco 2 nuclear reactor, proposed by the Sacramento Municipal Utility District for construction about twenty-five miles southeast of Sacramento, California.

Friends of the Earth nuclear expert James Harding scrutinized the company's cost projections and found the actual cost of power delivered by the Rancho Seco 2 reactor would be about three times more than the company's consultants, R. W. Beck and Associates, of Seattle, Washington, had predicted — in large part because the fuel-cycle costs would be five times higher than originally estimated.[49]

Harding's analysis drew a tacit endorsement from the Sacramento utility, which in 1976 decided to cancel the plant.

Utilities are now beginning to wake up and take note of the economic problems nuclear critics have attempted to alert the public about for years. Maybe the Northeast Utilities Co. will now heed the results of experts like Mr. Komanoff, for example, who testified in February 1976 at the Connecticut Public Utilities Control Authority's Generic Hearings on the Economics of Nuclear Power. Using standard nuclear economics references, including some highly conservative assumptions generated by the AEC, Komanoff was able to show that electric power from Northeast Utilities' proposed 1150 MW "Millstone 3" nuclear plant would not be cheaper than electricity from the twin 575 MW coal plants burning Appalachian low-sulfur coal that could be built in its stead.[50] Yet the coal plants present far fewer public safety problems.

A study on the future costs of coal power relative to nuclear power for the Canadian nuclear industry recently showed that if Canadian coal prices escalate at the rate expected for U.S. coal, and high capital-cost escalation continues, nuclear power will have no cost advantage in Canada for plants starting up in 1982. That study, "The Effect of Inflation on Nuclear Energy," was presented recently to the Canadian Nuclear Association by L. J. Schofield, manager of energy economics for Shell, Canada, Ltd.[51] It found that coal power would be cheaper for baseload generation if coal prices escalate at no more than 5 percent.*

Combustion Engineering, Inc. of Stamford, Connecticut, a firm which builds nuclear reactors and provides a variety of other energy-system design, engineering, and construction services, recently conducted a study of power plants scheduled for operation in 1980 and 1984. Although the company study found that nuclear power would be cheaper than oil and Eastern coal-fired plants with scrubbers, it concluded that in 1984, "A coal-fired system using western strip-mined coal requiring *no scrubbers* would have the lowest cost of all, 25.6 mills per kWh. And a nuclear system with a pressurized water reactor would produce electricity at a projected cost of 27.3 mills per kWh."[52] Thus it seems that even a company in the reactor business acknowledges that, at least in some instances, coal will be cheaper than nuclear power.

The preceding comparisons of nuclear and coal costs have focused on current and near-future dollar costs. But not all costs of energy technologies are fully reflected in the marketplace: neglected social costs, environmental effects, public health effects, safety risks, uncounted government subsidies, and costs of incipient technological "advances," such as the breeder reactor, tip the balance overwhelmingly against nuclear power, both now and in the foreseeable future.

THE BREEDER — LYNCHPIN OF THE NUCLEAR PROGRAM

Commercialization of the breeder, once slated for the 1980s, has now been postponed to the 1990s, and its costs are rapidly becoming prohibitive. Although in the late 1960s the AEC estimated the breeder's capital cost would be only $150/kW — 10 percent more

---

*To baseload a plant means to operate it as much of the time as possible, in contrast to load-following or operating the plant only to meet peak power demands. Because nuclear fuel has been cheaper than coal of oil so far, nuclear plants are typically used for baseload generation so as not to keep their huge capital investments idle any longer than necessary.

than a conventional reactor — the General Accounting Office recently estimated the cost of a 1000 MW breeder reactor to be 275 percent as much as an ordinary nuclear plant.[53] The prototype 350-400 MW demonstration breeder reactor now under construction on the Clinch River near Oak Ridge, Tennessee, has grown in cost from about $700 million to about $1.8 billion and may cost $2 billion before it is completed.[54] The other major breeder project in the U.S., the Fast Flux Test Reactor at Richland, Washington, once bore an $87 million price tag and is now pegged at $750 million.* Already $2.2 billion has been spent on breeder research and the total cost of the breeder program that the AEC in 1971 said would be $3.9 billion now is likely to cost $10 billion by 1982, with no commercial breeder in prospect for about a decade after that. Although breeder advocates point out that 40 percent of the $10 billion is attributable to inflation, the breeder costs still are enormous and that is only one of the reactor's serious problems.

The breeder's chief virtue is that it can be fueled mainly with cheap and abundant uranium 238 and that, thereafter, it ultimately produces a plutonium mass greater than the reactor's initial fissionable fuel charge. Theoretically, the excess plutonium would then be available for use as "mixed oxide" (uranium and plutonium) fuel in ordinary reactors. However, the breeder has a *long* fuel "doubling time."

Originally breeder enthusiasts asserted the reactor could double its fuel charge in only a few years and ERDA currently assumes the doubling time is fifteen years. As noted previously, the actual observed doubling time is much longer. This makes it all the more unlikely for the breeder ever to become commercially viable.

As design features are introduced to shorten the breeder's doubling time, the reactor becomes increasingly more dangerous. Yet even without such modifications, the breeder is inherently far more lethal and hazardous than a water-cooled reactor. A breeder contains more radioactive fuel in a more confined volume, and breeder reactions take place many thousands of times faster, than in conventional reactors, so safety devices have less time to work. Moreover, nuclear fuel can become supercritical and explode in a breeder during a fuel meltdown accident. Also, violent chemical

---

*Including the cost of component testing.

explosions can occur in a breeder because its liquid sodium metal coolant is highly explosive upon contact with water or air.

In addition, the copious production of bomb-grade plutonium in breeders intensifies both the threats of illicit atomic bombs and the dangers of environmental contamination from plutonium. These hazards make the breeder unlikely to be acceptable to the public. Finally, because its cost overruns are alienating its friends in Congress, the dollar-hungry breeder program probably faces substantial budget cuts. Yet if the breeder is *not* introduced to mitigate the fuel supply problem and more water-cooled reactors are built, nuclear fuel will surge upward in price again, eroding the one slim economic margin upon which all hopes for nuclear power's cost advantage once rested. Regardless of the final cost of building breeders, if they are to be commercially built, the $10 billion developmental cost is another form of subsidy to the conventional nuclear industry.

NUCLEAR SUBSIDIES

Subsidies, which disguise the true economic cost of an activity, are a venerable tradition for the nuclear industry. The industry was built on subsidies and, as we shall see, probably would not exist today without them. In the late 1950s and early 1960s, the AEC through its "Cooperative Reactor Demonstration Program" provided operating and construction subsidies for the first few commercial nuclear reactors. During 1963 to 1965, Westinghouse and General Electric sold several reactors to utilities at substantial losses so as to establish themselves in what then appeared to be a lucrative new field. These plants, known as "turnkey" units, were guaranteed by the reactor manufacturers to produce power for their purchasers at lower overall costs than fossil units. Many utilities were fooled into thinking nuclear power would stay cheap for them. Although the turnkey subsidy ended in 1965, commercial nuclear reactors retained their status as "experimental" devices until 1971, enabling them to continue absorbing AEC funds earmarked for research and development. Reactors have operated on a completely commercial basis for only five years.

From its inception, the commercial nuclear industry has also enjoyed an immense insurance subsidy from the federal government. In 1957, the year the country's first nominally commercial nuclear reactor went into operation at Shippingport, Pennsylvania, Congress passed a nuclear utility insurance law, the Price-Anderson Act, which became Section 170 of the Atomic Energy Act of 1954. This amendment was vital to the nuclear industry because it provided publicly subsidized insurance to nuclear plants that were largely uninsurable *and would not have been built without insurance.*

Specifically, the act set maximum liability limits for a nuclear power plant accident at $560 million. This limit means that total payments to compensate the victims of a nuclear accident can under no circumstances exceed $560 million, although the 1965 Brookhaven report update (see Chapter 3) had estimated that the worst conceivable accident could do $17 billion damage. Obviously this is not insurance in the conventional sense as it provides only pennies of compensation for every dollar of injury in the case of a nuclear catastrophe. But the U.S. insurance industry was not even willing to write policies to provide $560 million in coverage. The U.S. insurance industry, even after forming a "pool" of companies to spread the monetary risks of providing accident coverage, would offer individual utilities no more than $110 million in coverage — only 20 percent of the arbitrary federal liability limit.

Unwilling to expose utilities to the risk of bankruptcy in the event of a nuclear disaster, utility representatives testified, when Price-Anderson was introduced in 1957 and renewed in 1965, that they would not build nuclear plants if they had to bear the full liability. So in the same bold stroke with which Congress established liability limits, it authorized the AEC to sell utilities supplementary insurance up to $450 million, so they would be fully insured for the $560 million legal liability limit.

The passage of the liability limits and the supplementary insurance thus served as a decree establishing the nuclear industry. Today, nineteen years after the passage of the Price-Anderson Act, during which time the dollar has lost 50 percent of its purchasing power, the insurance industry is still willing to extend only $125 million coverage for nuclear accidents. Moreover, the premiums which the insurance industry is charging utilities for nuclear insurance, according to Senator Mike Gravel (D-Alaska), "suggest a risk-estimate (for a nuclear disaster) more like one-chance-in-twenty than one-per-billion,"[55]; the former estimate appears to be a less fanciful one.

In addition to helping the virtually uninsurable nuclear industry establish itself, the Price-Anderson Act has forced taxpayers to accept an insurance risk which private insurers deem actuarily unsound. And the arbitrary liability limits of Price-Anderson leave the lives, health, and property of Americans unprotected from large nuclear disasters. Nor do homeowners' insurance policies fill the liability gap, because they specifically exempt insurance companies from paying damage caused by nuclear accidents and radiation.

Without the protective mantle of Price-Anderson, utilities would either have to shut down nuclear plants for lack of adequate insurance, or they would have to try — at progressively greater and greater expense — to make those plants acceptably safe to the insurance industry.* Fear of this cost leads nuclear advocates into the inconsistent position of publicly claiming their plants are safe while resorting to the Price-Anderson Act because insurance companies believe the plants are too dangerous. That act has recently been renewed until August 1, 1987.

Government operation of uranium enrichment plants has concealed another kind of subsidy to nuclear power. The plants use enormous amounts of electricity derived from the burning of strip-mined coal in conventional power plants. By the 1960s, when these plants began providing substantial amounts of enriched uranium for commercial utilities instead of just for weapons, their capital costs had been absorbed by the U.S. government and they were drawing about 5 percent of all electricity used nationally but at low, federally-set bulk-rate prices, courtesy of the federally-owned Tennessee Valley Authority. In addition, the plants pay no taxes, insurance, or profits to investors. Consequently, the nuclear industry enjoys a lower price for its enriched uranium than would result from private facilities.

If nuclear power *really* held an overall economic advantage over coal and oil, then the nuclear industry should be willing and able to provide private entrepreneurs with an adequate rate of return for the provision of enrichment facilities *without* special government incentives. The Nuclear Fuel Assurance Act, the Administration's enrichment subsidy proposal, is simply a way to shift

---

*That could well be technically unfeasible or prohibitively expensive.

diseconomies of the nuclear industry onto the taxpayers to pay for the utility industry's commitment to an uneconomic technology. Such requests for subsidies by a supposedly commercial industry are a sign of economic failure.

The terms of the proposed subsidy reveal much about how the U.S. government coddles giant multinational corporations who deserve it least: ERDA would provide and guarantee its enrichment technology to private industry, and would provide backup stockpiles of uranium in case there were hitches in production. If the enrichment venture failed for any reason, that is, if it looked unprofitable to the contractor, ERDA would have to buy out the U.S.-owned 40 percent share of the project.

Other federal subsidy programs have benefited the nuclear industry by supporting research on nuclear plant safety, safeguards against thefts of bomb-grade nuclear materials, test reactors, new fuels, and waste management. Although waste management is an integral part of the nuclear fuel cycle, industry has been able to shift the cost of figuring out how to do it onto the taxpayers.

A careful look at cost projections for nuclear plants reveals yet another hidden subsidy. While building and operating plants, utilities have often conveniently ignored the eventual costs of entombing a reactor that is no longer serviceable, or restoring the radioactive reactor site to its original condition. Site restoration for a 1000 MW reactor would cost about $70 million (in 1982 dollars), according to James Harding of Friends of the Earth.[56] And if a reactor has to be "entombed," with its site kept fenced and guarded, the value of other foregone land uses and accrued taxes will be considerable.

Further subsidies for the entire utility industry have been formulated by the President's Labor Management Committee and endorsed by President Ford. These bailouts would increase some consumers' electricity bills and would certainly add to taxpayers' burdens. In mid-1975, the Committee made these recommendations:

(1) Permanently increase investment tax credits* for utilities to 12 percent, although they were earlier raised from 4 to 10 percent;

*An investment tax credit is a tax reduction granted to businesses. It allows them to reduce their taxes by an amount equal to a fixed percentage of their annual investments.

(2) Allow utilities to include construction work in progress in their rate base. This could increase the utilities' rate base on which public utility commissions compute consumers' utility charges;

(3) Extend special tax breaks for pollution-control investments mandated by law; and

(4) Defer taxes for investors in utility securities who reinvest in new company common stock.

These measures would certainly be an additional boost to an industry which has gotten more than three billion dollars in rate increases since 1973. The new subsidies would further obscure the cost of producing electric power and would tend to stimulate its consumption at a time when additional power consumption should be discouraged.

OTHER HIDDEN COSTS

In late 1975, President Ford proposed a $100 billion energy development corporation called the Energy Independence Authority to provide funds for capital-intensive energy projects during the next decade. Reintroduced into Congress in 1976, the giant corporation would use public money to advance nuclear and other expensive energy technologies with massive loans and loan guarantees. But the corporation's activities would not be reflected in the President's official budget, and its decisions would be subject to little public or congressional review.

In the nuclear area, the corporation's activities evidently could protect the troubled and noncompetitive nuclear power industry from market forces. Wielding open-ended subsidies of as yet undisclosed size, the corporation would have the ultimate subsidy device at its disposal: It could actually build nuclear power plants using lavish amounts of taxpayers' money and could later sell them at advantageous prices to utilities that initially could not afford them, or found them uncompetitive with coal plants. Not without reason did the *New York Times* in its editorial on September 23, 1975 dub Ford's plan, "essentially a bailout device for large corporations." The loser in this scheme once again is the U.S. taxpayer.

And so as to insure that energy consumers received the blessings of nuclear power as rapidly as possible, the national energy corporation also would be authorized to speed up regulatory agencies that might be delaying nuclear power plant construction. Thus, under color of promoting "energy independence," the new semi-autonomous corporation would have the funds to create nuclear hazards and the political clout to make the safety and environmental review process more perfunctory.

Although persuasive economic arguments can be made against nuclear power, economic cost ought not to be the definitive criterion in choosing an energy technology. Even if adding solar plants or clean coal plants with modern scrubbers to the nation's power grid were to cost more in a narrow economic sense over the short term, but without the hazards of plutonium, strontium, and cesium, it would *certainly* be worth the cost. And if we need governmental subsidies in the energy field, then let those subsidies be given to safe and less-polluting technologies, such as solar energy.

As complex judgments are made with tremendous human impacts, energy decisions begin leaving the realm of technology and enter the domain of social values and philosophy. As E. F. Schumacher pointed out so well in *Small is Beautiful,* the idea that something is "uneconomic" may simply mean that it is not monetarily profitable to the group of people who undertake it.

> Society . . . may decide to hang on to an activity or asset *for non-economic reasons* — social, aesthetic, moral, or political — but this does in no way alter its *uneconomic* character. The judgment of economics, in other words, is an extremely *fragmentary* judgment; out of the large number of aspects which in real life have to be seen and judged together before a decision can be taken, economics supplies only one — whether a thing yields a money profit *to those who undertake it* or not.[57]

It is precisely such a fragmentary judgment that has led us to adopt nuclear power as "a good buy," although it burdens us with large quantities of lethal materials and carries the risks of catastrophic accidents. But although radiation is harder to discern than the smoke and soot from coal, the monumental hidden and neglected costs of radiation are every bit as real as those of fossil-fired plants. If one arbitrarily assigns $300,000 as a compensation to the family of each of the eleven million people who could eventually die from radon-induced lung cancers, caused by uranium tailings, the dollar cost would be in the trillions, as extrapolations by David Comey of Dr. Robert Pohl's data show (see Chapter 5).

Radiation from nuclear reactors and the other phases of the fuel cycle are risks which we should never be compelled to take. Even assuming that the U.S. population receives only as much radiation as the maximum recommended by the Federal Radiation Council (170 millirads per year), tens of thousands of cancer deaths per year would result annually, as explained in Chapter 5. If the radiation dose to the U.S. population were only 100 millirads, the genetic health effects alone — mutations and deformities — would eventually cost $10 billion per year, over a period of generations, according to Dr. Joshua Lederberg, a Nobel Prize-winning geneticist. As for the other predictable health effects that will result from exposure of the general populace to legally allowed radiation levels, apparently no one has yet thoroughly tallied them and estimated their economic costs. Radiation, for example, might increase the incidence of stillbirths or cause subtle noncancerous thyroid disease or affect children's growth patterns.

Neither has the AEC or ERDA apparently done the requisite large-scale epidemiological studies of populations in the vicinity of nuclear power plants and their area-wide drainage systems to find out exactly what the effects of those plants have been so far. And whatever conclusion might be obtained from these studies concerning nuclear power plants in routine operation, those results would all be superseded in the event of a large-scale accidental release of radioactivity.

To see how much nuclear power really costs, it is necessary to total the costs of all its "routine" health effects now and in the future, its risks of catastrophic accidents and nuclear bomb proliferation, and its subsidies, including safety and safeguard research, fuel enrichment, waste management, and nuclear insurance. All these costs, when finally totaled and prorated to account for the tremendous "cost overruns" of large, low-capacity nuclear plants, would give an indication of how "cheap" nuclear power actually is. And when that final calculus is made, it would not be surprising if the partial moratorium on nuclear power imposed by soaring construction costs and high fuel costs were made final and total. The first step in this direction is to cut nuclear power off from all its insurance and federal operating subsidies, and then see how well it fares on a real-cost basis.

The folly of giving nuclear power a top billing among our energy research priorities despite its disappointing economic performance and health risks becomes clear by taking a glance at ERDA's direct energy research and development requests for fiscal 1976. Of a total $1.68 billion request, fission was the largest item, sopping up $763 million — or 45.5 percent of energy research and development funds. More than half this sum ($430 million) was requested for the breeder reactor boondoggle. By contrast, solar energy, which produces no radioactivity, ranks as such a low priority research item that ERDA requested only $89 million, or 5 percent, for it. That is just a third as much as for research on nuclear fusion. These priorities are even inconsistent with ERDA's own estimates that solar power can supply more electricity by 2000 than either the breeder reactor or controlled fusion.*

ERDA's apparent pessimism about the value of vigorously pursuing nonnuclear alternatives is not shared by many energy analysts, such as those experts at Tyco Laboratories of Waltham, Maryland, who expect to be able to produce electricity from solar cells more cheaply than from nuclear power plants. We will discuss their work in Chapter 11.

CONCLUSION

Today, the nation is in the absurd and vulnerable situation of dumping tremendous amounts of research and investment capital into a nearly worthless uranium fission economy. Only relatively small amounts of energy can be had from it in exchange for phenominal risks and costs. If we pursue this futile policy further, the whole country will be held as hostage for the profit of those nuclear interests that have speculated in uranium reserves and reactors. Further spending on uranium fission simply helps enrich these large energy firms which are not unhappy to see billions squandered on futile energy technologies instead of being invested in clean, productive, alternative energy sources.

---

*Controlled fusion is not even expected to be in commercial operation by then, and it may never be commercial. As yet, it has not been achieved for sufficient periods of time in a laboratory for it to be considered "scientifically proven." Once this is accomplished, its economic feasibility still would have to be demonstrated.

Table 7-4. AEC nuclear power projections

NUCLEAR ELECTRICAL CAPACITY IN GIGAWATTS

| | End of Calendar Year | | | | | |
| | December 1972 Forecast | | | | WASH–1139 (Rev. 1)[1] | |
	1980	1985	1990	2000	1980	1985
United States						
Most Likely _____	132	280	508	1200	151	306
High _____	144	332	602	1500	166	344
Low _____	127	256	412	825	132	272
Foreign (excluding Communist Bloc)						
Most Likely _____	141	303	578	1460	124	276
High _____	153	358	704	1900		
Low _____	123	256	454	1035		
Communist Bloc						
Most Likely _____	20	56	146	600		

[1] Interpolated from fiscal year data.

NOTES

1. *Wall Street Journal,* July 9, 1975.

2. Robert Gillette, "Atomic Energy Will Be A Prime Source of Debate," *New York Times,* February 9, 1975.

3. Ibid.

4. Daniel F. Ford, "Nuclear Power: Some Basic Economic Issues," Testimony Prepared for the "National Nuclear Energy Debate," Hearings Before the House Committee on Interior and Insular Affairs Subcommittee on Energy and the Environment (Cambridge, Massachusetts, April 28, 1975).

5. Irvin C. Bupp, et al., "The Economics of Nuclear Power," *Technology Review,* February 1975

6. "Monthly Electrical Equipment Update," White, Weld and Co., Inc., (March 11, 1975). White, Weld and Co., 1 Liberty Plaza, New York, NY 10006.

7. Leonard F. C. Reichle, "The Economics of Nuclear Power," presentation to the New York Society of Security Analysts, August 27, 1975.

8. James Harding, "The Deflation of Rancho Seco 2," *Not Man Apart,* v. 5, n. 15 (August 1975).

9. Reichle, op. cit.,

10. Nancy Ignatius and Joan Claybrook, eds., *Critical Mass '74 Handbook,* pp. 46-48. Center for the Study of Responsive Law, Box 19367, Washington D.C. 20036.

AN ESTIMATE of the ELECTRIC ENERGY DERIVABLE from URANIUM FISSION in LIGHT WATER ENVIRONMENT

(1) Uranium metal in U_3O_8
(2) Grams per short ton
(3) Fraction of U-235
(4) Fraction of U-235 entering cycle at 0.003 U-235 tails, 3.0% enrichment
(5) Probability of U-235 fission
(6) Burnup of U-235
(7) Thermal megawatt days per gram fissioned
(8) Thermal kilowatt hours per MWday
(9) Thermal-electric efficiency

A

$$0.848 \times 907200 \times 0.00711 \times 0.642 \times 0.855 \times 0.70 \times 0.92 \times 24{,}000 \times 0.33 = 15{,}353{,}387 \text{ kWhe/s.t.}U_3O_8$$

769,306 grams of uranium in 2,000 lb U_3O_8	5470 grams of U-235 per short ton of U_3O_8	3513 grams U-235 in cycle at 0.3% U-235 tails	3004 grams of U-235 subject to fission	2103 grams of U-235 actually fissioned	7302 kilowatt hours of electric energy per gram fissioned	Kilowatt hours of electric energy derived from fission of U-235 alone

B

FISSION OTHER THAN U-235:

Employing the empirical factor of 0.43 supplemental fissions per fission of U-235, 904 grams of material other than U-235 are fissioned for a gross energy yield of   6,603,126 kWhe

THE GROSS TOTAL ENERGY OUTPUT, adding A and B above, per short ton of U_3O_8,   21,956,513 kWhe

C

RESIDUAL FISSILE MATERIAL per short ton of U_3O_8:

Uranium-235:   3153 × 0.30 = 1054 grams

Plutonium-239: At 70 percent burnup of U-235, the residual Pu-239 may be significantly less than 400 grams, in which case, reprocessing of spent fuel would be pointless.

D

SIGNIFICANCE: All of America's uranium ore richer than 1500 ppm converted to electricity by this technology (one million tons U_3O_8 at 22 million kWh gross, 17 million kWh net) would supply one-fifth of the United States energy demand for five years.

Prepared by Morgan Gurdon Huntington and revised by him December 5, 1975

Chart 7-1. Estimate of electric energy derivable from uranium fission in LWRs

Appendix A   URANIUM FUEL DUTY

In deriving the number of reactors in the U.S. that can be fueled with our reserves of 690,000 tons of uranium,* we have used Huntington's figure of approximately 22 million kWh(e) for the gross energy yield per ton of U_3O_8. We have also assumed that all reactors are 1000 MW(e) in size and that they will be operated at a generously estimated 70 percent of capacity. This is 15 percent above presently observed capacity factors, but well within the range which the nuclear industry regards as practically attainable.

Using the data above, we find that each plant produces 245 billion kWh(e) during its lifetime:

$$1000MW(e) \quad x \quad .70 \quad x \quad 8.76x10^3 \quad x \quad 40 = 245x10^9 kWh(e)$$

| | Capacity factor | hours in a year | years lifetime | total power per plant |

Since each ton of fuel yields 22 million kWh(e) (i.e. $2.2x10^7 kWh(e)$), division of $\dfrac{245x10^9 kWh(e)}{2.2x10^7 kWh(e)/ton}$ yields 11,150 tons, the amount of fuel required by each reactor during its assumed forty-year lifetime.

With total fuel supply reserves of 690,000 tons, and each reactor consuming 11,150 tons, it is clear that only $\dfrac{690,000 \text{ tons}}{11,150 \text{ tons/reactor}} = 61.9$ reactors can be fueled throughout their lifetime.

If spent fuel is reprocessed to gain 8 million kWh(e) per ton of fuel, the number of reactors that can be fueled may be found by using the above calculation, but with 30 million kWh(e) per ton instead of 22 million kWh(e) per ton.

To calculate the net energy yield from a reactor, subtract 3.5 million kWh from the assumed energy yield per ton of fuel: with reprocessing this equals 26.5 million kWh; without reprocessing it equals 18.5 million kWh.

*   *   *

Note the assumptions made on the Huntington chart of uranium fuel duty:
- A 33 per cent loss of uranium-235 (left in tails) in the enrichment process.
- A 15 per cent loss of uranium-235 in the reactor by neutron-capture, producing unfissionable uranium-236 or uranium-237.
- A 43 per cent energy bonus per uranium-235 fission, mostly from the fissioning of plutonium-239.
- A 70 per cent fuel burn-up in every reactor fueling cycle.

---

*In the spring of 1976 ERDA revised its estimate of reserves recoverable at $30 per pound downward to 640,000 tons. Only 57 reactors can be fully fueled with 640,000 tons.

11. Ford, op. cit.

12. "Nuclear Power Saved $810 Million in U.S. Electric Costs in 1974," Atomic Industrial Forum press release, April 2, 1975.

13. "Fossil Generation Costs Nearly Double Nuclear's for First Half 1975," *Energy Finance Week*, v. 1, n. 20 (September 17, 1975), p. 6. Energy Finance Week, 1243 National Press Building, Washington, D.C. 20045.

14. Proponents of nuclear power like the Atomic Industrial Forum have frequently clouded the performance issue by using the terms, "availability" and "reliability" in describing plant performance. Availability, however, simply tells how much of the year a plant is *capable* of producing power, not how much power the plant actually produces, and therefore does not account for frequent periods in which plants can produce some electricity but cannot operate at full power. Thus a plant could be available 80 percent of the year but it might be restricted to producing only 40 percent of its rated capacity. Although the plant's total *availability* figure would be high (80 percent), its *capacity factor* would be only 32 percent (.4 x .8). By contrast to both availability and capacity, reliability is a value judgement about a power plant and is not pegged to any scientifically accepted measure for comparing plant efficiencies.

15. Reichle, op. cit.

16. "Westinghouse Eyes Loss on Uranium Contracts," *Energy Finance Week*, v. 1, n. 12 (July 23, 1975).

17. U.S. Energy Research and Development Administration, *Liquid Metal Fast Breeder Reactor Program, Final Environmental Statement* (ERDA - 1535), v. 1, p. III E-7, December 1975.

18. Ibid., pp. III E-4 to III E-10.

19. Ibid.

20. Morgan G. Huntington, "How Good Are Our Energy Reserve Estimates? How Much of this 'Energy Reserve' Can Be Made Available to the Economy as Net Usable Energy?" Monograph, 1976. (Woodfield Road, Galesville, Md. 20765)

21. Ibid.

22. Charles Komanoff, "Testimony Before Connecticut Public Utilities Control Authority in Generic Hearings on the Economics of Nuclear Power," Council on Economic Priorities, 84 Fifth Avenue, New York, NY 10011, Feb. 27, 1976.

23. Ralph E. Lapp, "We May Find Ourselves Short of Uranium, Too," *Fortune*, October 1975.

24. Huntington, op. cit.

25. U.S. ERDA, op. cit., p. IIIE-1.

26. U.S. ERDA, *Liquid Metal Fast Breeder Reactor Program, Proposed Final Environmental Statement* (ERDA - 1535), p. VV 55-20, December 1974.

27. John Klemenic, "An Estimate of the Economics of Uranium Concentrate Production from Low Grade Sources," Monograph, Planning and Analysis Division, Grand Junction Office, U.S. AEC, Grand Junction, Co., Oct. 22, 1974.

28. Carl L. Bieniewski, Franklin H. Persse, and Earl F. Brauch, *Availability of Uranium at Various Prices from Resources in the United States,* Information Circular 8501, Bureau of Mines, U.S. Dept. of the Interior, 1971.

29. Chauncey Starr, "Energy and Power," *Energy and Power* (San Francisco: W. H. Freeman and Co., 1971).

30. Lapp, op. cit.

31. Les Amis de la Terre, *L'Escroquerie Nucleaire,* Lutter/Stock No. 2 (Paris: Editions Stock, 1975).

32. Robert W. Fri (ERDA Deputy Administrator), speech prepared for delivery before World Fuel Market, Washington, D.C., Sept. 23, 1975. *Information from ERDA,* v. 1, n. 28, October 8, 1975, Washington, D.C. 20545.

33. Jonathan Kwitny, "Enriching Venture, How Firm Got Ahead in Billion-Dollar Race to Make Nuclear Fuel," *Wall Street Journal,* Nov. 20, 1975.

34. Ibid.

35. Bertram Wolfe and Ray W. Lambert, "The Back-End of the Fuel Cycle," presentation to the Atomic Industrial Forum, March 20, 1975.

36. Marvin Resnikoff, Professor of Physics, Rachel Carson College, State University of New York at Buffalo, Amherst, New York.

37. "Udall Hearings Explore Nuclear Power," F.A.S. Public Interest Report, v. 28, n. 5-6 (May-June 1975). Federation of American Scientists, 307 Massachusetts Avenue, N.E., Washington, D.C. 20005.

38. "Why Atomic Power Dims," *Business Week,* November 17, 1975.

39. Customary measures of capacity to date have not taken account of the fact that some plant capacity factors may be artificially lowered because the plant is used for load following and its power is not demanded for part of the measured operating period.

40. *Nuclear Power Plant Availability and Capacity Statistics for 1973,* Report 003-08-002, Office of Operations Evaluation, U.S. Atomic Energy Commission, Washington, D.C. (1974).

41. David Dinsmore Comey, "Will Idle Capacity Kill Nuclear Power," *Bulletin of the Atomic Scientists,* v. 30, n. 9 (Nov. 1974);"Nuclear Power Plant Reliability, the 1973-74 Record," *Not Man Apart,* v. 5 (mid-April 1975); "Future Performance of Large Nuclear Plants," *Not Man Apart,* v. 5, n. 14 (mid-July 1975); "No Improvement, Capacity Factors Stay Constant in 1975," *Not Man Apart,* v. 6, n. 4 (March 1976).

42. U.S. AEC, Office of Planning and Analysis, *Nuclear Power Growth 1974-2000* (WASH - 1139(74)), Washington, D.C., 1974.

43. Peter Margen and Soren Lindhe, "The Capacity of Nuclear Power Plants," *Bulletin of the Atomic Scientists,* October 1975, pp. 38-40.

44. Comey, "Nuclear Power Plant Reliability, the 1973-74 Record," loc. cit.

45. Komanoff, op. cit.

46. Con Edison news release, January 20, 1975.

47. *Customer News,* Con Edison pamphlet, April 1975.

48. Charles Komanoff, "Responding to Con Edison: An Analysis of the 1974 Costs of Indian Point and Alternatives," Council on Economic Priorities, 1975.

49. Harding, op. cit.

50. Komanoff, "Testimony . . . ," op. cit.

51. "Canadian Analysis Shows Low-Cost Coal May Undercut Nuclear in the Out Years," *Energy Finance Week,* v. 1, n. 12 (July 23, 1975).

52. *Progress,* n. 2, 1975. A publication of Combustion Engineering Inc., 900 Long Ridge Road, Stamford, Connecticut.

53. "Udall Hearings," loc. cit.

54. Allen L. Hammond, "Complications Indicated for the Breeder," *Science,* v. 185 (August 30, 1974); "Morton and Zarb Join in Suggesting a Slowdown on Nuclear Breeder Reactors and Call for More Research," *New York Times,* June 10, 1975.

55. Senator Mike Gravel, Testimony to the Joint Committee on Atomic Energy on S. 3254 and H.R. 14408, May 29, 1974.

56. Harding, op. cit.

57. E. F. Schumacher, *Small Is Beautiful* (New York: Harper and Row, 1973).

General References

Ralph M. Rotty, A. M. Perry, and David B. Reister, *Net Energy from Nuclear Power,* Institute for Energy Analysis, Oak Ridge Associate Universities, IEA-75-3, IEA Report, November 1975.

Michael Rieber and Ronald Halcrow, *Nuclear Power to 1985: Possible versus Optimistic Estimates,* CAC Document No. 137P, Center for Advanced Computation, University of Illinois at Urbana-Champagne, Urb ana, Illinois 61801.

Dennis Holliday and Vincent Taylor, *The Uncertain Future of Nuclear Power,* California Arms Control & Foreign Policy Seminar, P.O. Box 925, Santa Monica, Calif. Revised August 1975.

Frederick Q. Barnes, "Uranium — Where are the Reserves?" *Mining Congress Journal,* v. 62, n. 2, February 1976 (1100 Ring Bldg., Washington, D.C. 20036).

8

ATOMIC POWER PLAYS

Nuclear energy in the United States has been championed by a powerful and pervasive coalition that has been called the "Atomic Industrial Complex." To understand the incredible story of how a fundamentally uneconomic technology has spread so rapidly in the U.S. and abroad, we must analyze the make-up of this coalition. The forces behind it are a formidable amalgam of industrial, governmental, corporate, and academic interests that are the political muscle behind nuclear power. Although this nuclear lobby has a preeminent impact on American energy policy, the general public knows little of this lobby's political and economic influence. Perhaps this is because, until recently, the nuclear legions met, in general, with only slight resistance.

For years, the nuclear critics' ranks were thin. They included only a few poorly funded environmental organizations, grassroots citizens' groups, and professional public-interest research outfits. These groups today are aided by a growing number of rank-and-file supporters, and by a smattering of academic analysts, attorneys, nuclear industry whistle-blowers, and maverick scientists. Arrayed against this fragile alliance is a well-entrenched, multibillion-dollar complex of national, multinational, and regional corporations, united in an unseemly compact with important government officials and agencies.

The complex consists of huge oil companies, leading uranium mining firms, major utilities, enormous financial institutions, large defense contractors, top nuclear equipment manufacturers, vast construction companies, big architect-engineering firms, and all their respective subcontractors, and hired industry pressure groups. The latter carry the nuclear industry's case directly to every level of government. In this fashion, the nuclear complex pervades the entire economy and influences the highest echelons of government.

The pronuclear forces include firms such as Exxon Corp. (formerly Standard Oil of New Jersey) and Gulf Oil Corp. Both have larger revenues than many foreign nations, such as Colombia or New Zealand. Standard was founded by John D. Rockefeller (Sr.) and Rockefeller interests have major holdings in the firm. The company had total revenues of about $45 billion in 1975. The firm owns uranium properties, and Exxon Nuclear produces nuclear fuels. Smaller oil companies like Continental Oil Co. and Kerr-McGee Corp. hold valuable uranium acreage. Kerr-McGee is the largest U. S. producer of uranium, and has an interlocking directorate with General Electric Corporation.* An interlock occurs when a director of one company also sits on the board of directors of another. GE has a direct interlock with Utah International Resources (formerly Utah Construction and Mining Co.), which annually produces millions of pounds of uranium oxide. Indirect interlocks connect GE to Rockefeller Bros., Inc. Gulf Oil Co., in turn, with total revenues of $17 billion, has interlocks with Westinghouse Electric Corp. and the Mellon National Bank, among other corporations. A director of Westinghouse is also on the Mellon board; another directs the Pittsburgh National Bank.

Interrelationships between firms in the energy industry are extremely significant because they can provide clues to possible

---

*Information on interlocks in the energy industry comes from two major sources:
1) Arthur D. Little Co., *Competition in the Nuclear Power Supply Industry*, Report to USAEC and US Dept. of Justice, Contract No. AT (30-1)-3853, Dec. 1968. See figure B-1, "Corporate Relationships Through Director Affiliations."
2) James D. Sneddon, "The Nuclear Web - Who Benefits From the Pursuit of Nuclear Power," *Beaver County (Pa.) Times*, Dec. 24, 1974. The same issue contains five other valuable major features on nuclear power, resulting from months of independent investigation by Sneddon.
Because time has passed since this information on interlocks was collected, changes may have occurred in the compositions of some boards of directors mentioned. Nonetheless, the basic interlock matrix described here is still relevant and useful.

conflicts of interest within the nuclear complex.* Such conflicts suggest possible reasons why a utility might opt for nuclear plants when other power plants are cheaper, safer, and more reliable. The decision to buy the nuclear capacity might result from utility executives' mistaken expectations that the nuclear plant would have a lower lifetime cost. But if utility directors have vested interests in other interlocked firms that benefit from the utility's nuclear expansion, such benefits could accrue to the firms even if the nuclear commitment ultimately hurt the utility actually making it.

The nuclear complex benefits from the prevalence of industrial interlocks and from the large investments required for a nuclear plant. Major conflicts of interest can occur in the letting of contracts to construction firms, steel companies, banks, and other firms. Meanwhile, the decision by the utility to "go nuclear" — even if unsound in the long run, initially tends to cause the value of the utility's stock to rise (by expanding the company's investment capital, thereby increasing the company's value and its rate base†). Needless to say, when a utility's stock appreciates, utility directors are often among those who benefit through their stock holdings.

Banks are naturally disposed to look kindly upon large construction projects in general, because they require substantial amounts of money and often necessitate large loans. This is especially true in the case of nuclear power plant construction, which requires far more capital investment than the construction of alternative generating stations burning coal, oil, or natural gas.

Many banks own utility stocks through the banks' trust departments, and banks, therefore, can vote in favor of nuclear commitments at stockholders' meetings. Tracing connections be-

---

*Explanations of research methods useful for probing the economic ties of nuclear proponents can be found in the Corporate Action Project's Corporate Action Guide (Washington, D.C.: Corporate Action Project, 1974). CAP, 1500 Farragut St., N.W., Washington, D.C. 20011. See also *NACLA Research Methodology Guide* (North American Congress on Latin America: New York, 1970). NACLA, Box 226, Berkeley, Calif. 94701.

†The rate base is a measure of the amount of money a utility has invested to do business. It represents the current value of a company's assets — generating plants, office buildings, transmission lines, vehicles, tools, and supplies — needed to sell power. The electricity rates a company is allowed to charge its customers are calculated by the Public Utilities Commission so as to insure the company a return equal to a fixed percentage of its rate base.

tween utilities and their banks can reveal another mechanism by which special interests may be served by nuclear plant construction, regardless of the soundness of that investment. The bank may incur a conflict of interest by (1) owning part of the utility (through stock holdings) and then, (2) loaning it money for power plant construction through the bank's loan department. Further conflict could occur if the bank also provides underwriting or other services for the utility's stocks and bonds. For example, banks can make large commissions by serving as utility bond trustees or escrow agents (which often hold millions of dollars until the funds are actually needed for plant construction).

The nuclear complex includes many of the country's approximately two hundred investor-owned electric power utilities, such as Consolidated Edison of New York, Wisconsin Electric Power Company, and Pacific Gas and Electric Company of California, as well as the publicly owned Tennessee Valley Authority, which currently plans to build 20 new nuclear plants. Typically, the investor-owned pro-nuclear utilities are backed by major banks, like Bankers Trust Company of New York, Chase Manhattan Bank of New York, and New England Merchants National Bank. Chase Manhattan is closely tied to the Rockefeller Family through the presidency of David Rockefeller, and by various interlocks. This bank is among the ten largest security holders of 42 utilities, according to the Senate Committee on Government Operation's recent *Disclosure of Corporate Ownership.* Morgan Guaranty Trust, Manufacturer's Hanover Trust, and First National City Bank were among the top ten investors for scores of other utilities.

Major insurance corporations, mutual funds, foundations, and universities also invest heavily in the country's utilities and in other large energy firms.

Utilities are often attractive investments. They operate in a legalized monopoly environment and have assured rates of profit, set by public utility commissions. In addition, should fuel costs rise, most electric utilities are allowed to automatically "pass through" their increased costs to their customers.

Interlocks are plentiful not only within the energy complex, but also from nuclear companies to firms that do not at first appear to be implicated in energy investment. Equitable Life Assurance Society of the U.S., for example, interlocks with United Nuclear Corp., a uranium mining firm, and with Combustion Engineering, a reactor manufacturer. Both United Nuclear Corp. and Combustion Engineering interlock with Rockefeller Bros., Inc. Tracing Equitable Life's connections further shows how deeply nuclear interests permeate the economy. Equitable is interlocked not only with Detroit Edison Co. but with American Telephone and Telegraph Co., Proctor and Gamble Co., and the Massachusetts Institute of Technology. Proctor and Gamble interlock with Gulf Oil, MIT with Westinghouse. And those are but the first few strands in the complex web that binds nuclear interests into the financial community.

In addition to their interlocks, the financial sponsors of the nuclear lobby are also economically tied, through loans, and stock/bond ownerships, to major U. S. industries, including the major nuclear-reactor makers: General Electric Co., Westinghouse Electric Corp., General Atomic Co., Babcock & Wilcox Co., and Combustion Engineering, Inc.*

The financial backers of the energy corporations are, of course, linked to multinational construction giants like Bechtel Corp., or to Brown and Root, Inc., both builders of nuclear plants and other projects. Numerous smaller architect-engineering consulting firms and their subcontractors also have stakes in nuclear construction. Stone and Webster, Inc., a major architect-engineer, has an interlocking directorate with Chase Manhattan Bank, for example. Ebasco Services, Inc., another nuclear architect, interlocks with the Manufacturers Hanover Trust Co. through Ebasco's parent firm, the Electric Bond and Share Co. Like Brown and Root, Ebasco is a subsidiary of Halliburton Co. All these large companies can profit greatly from nuclear power expansion regardless of whether or not it is in the public interest.

Large numbers of scientists and engineers are directly employed by the nuclear industry, or perform research, consultations, or academic assignments on contracts predicated on the industry's viability. They are on the industry's payroll and they form an important phalanx of the nuclear lobby. Their technical papers and reports in the nation's leading scientific journals and in subsidized nuclear industry publications provide a cascade of pronuclear research findings with which to justify further nuclear endeavors and further research grants.

---

*Allis-Chalmers and Tenneco Oil Co. are new entrants to the reactor field.

The record is clear on how this situation developed. For 28 years, the biggest nuclear proponent of all time, the Atomic Energy Commission, provided the research grants, contracts, or salaries for the vast majority of all the nuclear scientists and engineers in the country. It also created a phenomenal deluge of free pro-nuclear literature and audiovisuals, much of which, in all candor, can only be called propaganda.

In addition to the nuclear juggernaut's friends in organized science, powerful federal and state agencies have also allied themselves with the nuclear industry. The Ford Administration and Frank Zarb, head of the Federal Energy Administration have strongly supported rapid nuclear expansion. The U.S. Energy Research and Development Administration, comprised largely of the Atomic Energy Commission minus its regulatory division (now the U. S. Nuclear Regulatory Commission) is a stalwart supporter of nuclear fission, as even a cursory look at ERDA's budget reveals. The U.S. Environmental Protection Agency has staffers in its radiation division who were formerly employed by the AEC.

In Congress, the 18-member Joint Committee on Atomic Energy has dominated all nuclear legislation since 1946, successfully stifling legislative criticism of nuclear power and moratorium efforts. It is the only joint committee created by statute, and the only committee with authority to originate legislation. The committee also has the extra power to introduce its bills simultaneously in both houses of Congress, and from 1963-1974, has had the power of reviewing the AEC's budget. During most of its lifetime, the committee has been heavily influenced if not dominated by the pronuclear advocacy of its now-retired senior spokesman, former Congressmen Chet Holifield (D-Calif.), and Craig Hosmer (R-Calif.).

NUCLEAR PRESSURE GROUPS

In addition to the active public relation departments of major utilities, many trade, scientific, and public relations associations are vigorously promoting nuclear power. Prominent among these are the Edison Electric Institute, the Atomic Industrial Forum, the American Nuclear Society, the American Public Power Association, and Reddy Kilowatt, Inc. of Greenwich, Connecticut.

The American Nuclear Society of Hinesdale, Illinois is the

nuclear industry's principal professional organization, with a membership of 10,000 scientists, engineers, and administrators professionally engaged in nuclear science or engineering. Its avowed purpose is to promote the advancement of research and operations in fusion, fission power, and non-conventional fuels technology. In addition to publishing four important nuclear periodicals, the society modulates the flow of information on nuclear power by offering energy-related information services, a speakers' bureau, as well as meetings that enable nuclear proponents to coordinate their activities. The society's November 1975 convention was held jointly in San Francisco with the Atomic Industrial Forum.

Established in 1953, the Atomic Industrial Forum, with headquarters in New York, is a primary nuclear industry pressure group. Among its 600 member organizations can be found a wide variety of industrial groups, research and service organizations, educational institutions, labor groups, and government agencies concerned with nuclear power.

The forum promotes public acceptance of nuclear power through a large variety of information services. Its monthly newsletter, *Press INFO*, reaches 1,500 editors and reporters as well as 400 representatives of state regulatory agencies and special interest groups. The AIF publishes and distributes everything from technical books and popular pamphlets, to scientific studies and news releases on nuclear power, and it offers an annual award "for public understanding of nuclear energy by news media." Nuclear critics have yet to match this systematic media barrage.

The forum will also perform research services for people writing on nuclear power. Authors writing books on the subject will find the forum eager to "get involved" by reviewing the manuscript. Pronuclear films and nuclear graphics can be borrowed without charge from the forum, and the organization maintains a 7,000-volume library and a speakers' bureau.

Other functions of the Atomic Industrial Forum include representing U.S. and foreign nuclear companies in government and public affairs; acting as a liaison between the industry and government; providing input to an AEC reactor safety study; and serving as a clearinghouse on nuclear industry problems, such as those regarding plant licensing, siting, construction, safety considerations, etc.[1] The forum also regards itself as a research group and compiler of nuclear statistics. The 1974 edition of *The Energy Directory* did not give the Forum's exact budget, noting only that it was "over $1,000,000." This statement is rather disingenuous in that the forum's proposed Public Affairs and Information Program budget alone for 1975 was $1.3 million.

Another vocal advocate of nuclear power is the Edison Electric Institute (EEI), the national coordinating group for U.S. investor-owned utilities and electric utility holding companies. Founded in 1933, the EEI now has 226 utility members. The institute is an essential research and public relations arm of the utility industry, with divisions of economics and administration; engineering and operating; industrial relations; public and investor relations; research; publications; and statistical reports.

It conducts research and development on breeder reactors, fusion, clean-coal technologies, electrical transmission, and electrical storage. EEI publications include weekly, bimonthly, semiannual, and annual statistical surveys of the electric utility industry, which are widely used as references. Some of the EEI's research is done in collaboration with the federal government and nuclear equipment manufacturers. The institute also reviews those plans of members that have public impacts, and may coordinate advertising campaigns.

Publicly owned utilities and electric utility cooperatives are coordinated by the American Public Power Association of Washington, D.C. The APPA conducts research and development and provides information.

An important industry association with a promotional orientation toward electric power is the Electric Energy Association of New York City. The Association assists investor-owned utilities and the manufacturers of electrical heating equipment and insulation. One of its chief tasks at a time of energy shortages is to promote electric heating systems in the design and construction of buildings, by disseminating information to engineers and architects. Electric heating systems are inefficient converters of fuel to heat, compared with on-site oil or gas heating units.

A recent arrival in the ranks of the in-house industry research groups is the Electric Power Research Institute, with offices in Palo Alto, California, and Washington D.C. Although founded in 1973, EPRI is of paramount importance to nuclear proponents.

Come live in the electric climate.

The electric climate can do more than help give you more room for living...

It's a superior indoor environment that can make any kind of building cost less to own...and it can help the outdoor environment, too!

Consider what the benefits of *the electric climate* can mean to you as a homeowner. And as a cost-conscious, people-conscious executive. And as a civic-minded citizen.

The human benefits of the electric climate:
Flameless electric heat is the heart of *the electric climate*. It fills rooms with a soft, even warmth that can't be matched for comfort. No drafty corners. No sudden chills. Except for the comfort, you hardly know it's

there–whether you're in your electrically heated home or office or church or school. Think how much better people live and play and learn and work in such a pleasant environment.

The dollar value benefits of the electric climate:
The initial cost of flameless electric equipment that results in *the electric climate* is comparable to or *lower* than other types. Requires little or no maintenance. And the *cost* of electricity remains a real bargain!

The environmental benefits of the electric climate:
Buildings with *the electric climate* use the cleanest source of energy there is. *Flameless* electricity. There's no combustion! Therefore, buildings with *the electric climate* put nothing in the air around them!

The *electric climate* isn't a promise for the future. It's here *now.* Find out more about it from your electric utility company. You'll benefit. So will your company. And your community.

Live better electrically / Move toward a better world.

Edison Electric Institute, 750 Third Avenue, New York, N.Y. 10017

Figure 8-1. Electric power promotion

Participating electric utilities pay a tithe to EPRI for every mill of electricity they sell. The EPRI budget amounts to $68,000,000 annually for conducting research in all phases of electricity generation and transmission. The organization had no less than 450 research projects under its purview or in contract negotiations as of February 1, 1975. These projects will cost over $300,000,000 during their contract lifetimes, including funding from other organizations[2] EPRI engages in cooperative research agreements with the Energy Research and Development Administration, the Office of Research and Development of the Department of the Interior, and the National Aeronautics and Space Administration. Among EPRI's highest priority short-range goals for the 1975-1985 period is to "Strengthen the nuclear power option and resolve nuclear safety issues, with the objective of reducing the time required to obtain site approval and to construct nuclear power plants."[3]

Sometimes it is to the advantage of electric utilities and their associations to set up or support second-generation organizations or shadow organizations to further the parent group's interests. For example, the Edison Electric Institute and several electric utilities provided support to an organization called the Federation of Americans Supporting Science and Technology (FASST). Two important goals of FASST in 1974 were to establish Energy Youth Councils on college campuses throughout the U.S. and to funnel news to 1,000 college newspapers through FASST News Service. This mass production and shotgun distribution of premasticated "news" from the energy combine goes out with the apparent blessings of some pivotal people in government: FASST's advisory board includes Senator Henry M. Jackson, Senator Frank E. Moss, and Congressman Mike McCormack. The obvious aim of enlisting college students on energy councils is to create organizational structures that can be used to muster campus support for the atomic complex or to create the premature impression that such support has already solidified.

Some utilities employ the services of an agency calling itself Reddy Kilowatt Inc. of Greenwich, Connecticut. Through its Fact Systems division, Reddy Kilowatt distributes pronuclear literature such as *The Nuclear Controversy,* by Dr. Ralph Lapp. They also have furnished reams of pronuclear literature to highly placed, senior electric-utility executives. Reddy Kilowatt performs other important chores for the nuclear advocates. For example, the Fact Systems division scheduled a "combat training" seminar in June 1974 to teach utility executives how to cope with the sometimes emotional opposition to nuclear power they are liable to encounter. The program featured staged debates on nuclear issues which Fact Systems videotaped and showed to participants. According to a Fact spokesperson many utility presidents and vice-presidents agreed to participate in the three-day session at a cost of $800 per person. The executives also were invited to briefings by nuclear advocates, such as Dr. Lapp.

Some nuclear proponents retain the services of a public relations-advertising firm, such as Underwood, Jordan Associates of New York City, which operates an "Electric Companies Public Information Program." One of its functions is to keep its clients apprised of nuclear critics' activities. A casual review of their newsletter indicates the critics' political activities and literature are diligently monitored.

Another recently founded second-generation pressure organization is the American Nuclear Energy Council, headed by Craig Hosmer, former Chairman of the Joint Atomic Energy Committee. Supported by nuclear utilities, reactor manufacturers, and nuclear architect-engineering firms, the council's main function is to lobby for nuclear power. Its offices are next door to those of the Nuclear Regulatory Commission and many of ANEC's members also belong to the Atomic Industrial Forum.

Still another upper echelon lobbying effort that also set up shop in 1975 within striking distance of Capitol Hill is Americans for Energy Independence. The group had received $200,000 from General Electric Co., Westinghouse Electric Corporation, and major utilities to marshall the forces of organized labor behind the nuclear energy bandwagon. AIE is chaired by physicist Hans Bethe, a Nobel laureate. Former AEC Chairman Dixy Lee Ray is to be on the board of directors.

Two other significant new nuclear pressure groups that are playing major roles in committing the nation to a nuclear future are the California-based Citizens for Jobs and Energy and the California Council for Environmental and Economic Balance. Both were active in mustering public opposition to the California Nuclear Safeguards Initiative, a ballot proposal much feared by the nuclear industry because it places environmental and safety restrictions on the unbridled growth of the Atomic Industrial Complex.

THE CALIFORNIA NUCLEAR INITIATIVE

As this book goes to press, the initiative will be decided by California voters at the polls in June 1976, as a result of a grassroots petition signature drive conducted during 1975. If voters approve it, the initiative could put a stop to the construction of all new nuclear power plants in California and require that existing plants be operated at greatly reduced fractions of their design capacities, thereby placing a stiff economic penalty on their use. New plants would be prohibited — and existing plants' power output would be restricted — unless Californians were assured full insurance coverage for the damage in a major nuclear accident. Existing plants also would be restricted in their output unless emergency and other safety systems were subjected to comprehensive, full-scale tests.*

The roots of the two important organizations opposing the initiative run deep into the substrate of California politics. Former California Governor Edmund G. (Pat) Brown (Sr.) is chairman of the anti-initiative California Council for Environmental and Economic Balance. His political connections and savvy after years in the state capitol must have been an asset when, early in the anti-initiative fight, he reportedly contacted all the members of the California legislature, urging them to oppose the initiative. The group is nominally directed by Michael Peevey, a relatively un-

---

*The Nuclear Safeguards Initiative, legally known as Title 7.8 LAND USE, NUCLEAR POWER LIABILITY & SAFEGUARDS ACT, is proposed as an amendment to the State of California Government Code. The measure provides that one year after its passage, no new nuclear plants may be built in California, and existing plants must be derated to 60% of their licensed capacity *unless* the current federal insurance liability limit ($560 million) for nuclear accidents has been removed. The initiative also stipulates that five years after its passage, no nuclear plants shall be operated at more than 60% and must thereafter be derated at 10% of their licensed capacity per year *unless* the liability condition is met and nuclear power plant safety and waste storage systems have been demonstrated to the satisfaction of the California legislature. The legislature's approval of safety and storage systems must be by a two-thirds majority. The complete text of the initiative is given in Appendix B.

known administrator, formerly with the U.S. Department of Labor. The Brown-Peevey group has been raising money and support for the anti-initiative fight, coordinating their activities with Citizens for Jobs and Energy, of which Brown is a co-chairman.

The seven other co-chairpeople of Citizens for Jobs and Energy (CJE) include the former general manager of the Los Angeles Department of Water and Power as well as representatives from organized labor and influential people in the business community. CJE's comptroller is a former member of the state's Public Utilities Commission, and the group is supported by a forty-member organizing committee.

CJE's activities are coordinated by the public relations firm of Winner-Wagner & Associates. Charles Winner is active in Democratic Party politics and has assisted Senator John Tunney (D-Calif.) and former California Assembly Speaker Bob Moretti, who sits on the state's five-member Energy Resources Conservation and Development Commission.

Various engineering societies have also lent their support to the anti-initiative movement. The 170,000-member Institute of Electrical and Electronic Engineers opposed the initiative, and 2,600 members of the Power Engineering Society voluntarily donated thousands to the same cause. The Los Angeles chapter of the American Nuclear Society also tentatively approved a full-blown public relations and lobbying effort against the initiative.

The pro-initiative forces consist of citizens and environmental groups such as the People's Lobby of Los Angeles, Project Survival of Palo Alto, the Sierra Club, Friends of the Earth, and the California Citizen Action Group, a branch of the Ralph Nader organization. Richard Spohn, a lawyer and chairman of Citizen Action, heads the initiative campaign with lawyer David Pesonen of Berkeley and others.

Currently, it is too soon to know exactly how the nuclear proponents, orchestrated by Winner-Wagner & Associates, will conduct the home stretch of the anti-initiative campaign. But it is already obvious that the big corporations, big money, and the ranks of the scientific establishment are committed to the initiative's defeat. Whereas the supporters of the initiative have received no corporate contributions, the opponents have raised contributions in the four and five figures.* A memo from Bechtel Corporation's Paul W. Cane to the company's Board of Directors detailed the steps the firm plans to take in trying to defeat the initiative. Naturally, many company workers will vote on the initiative, and labor unions in California, too, have been encouraged by the initiative opposition to conduct "educational campaigns" for their members.

Some opponents of the initiative are currently circulating a counter petition to qualify a Domestic Energy Initiative for California ballots in November 1976. This new initiative could nullify the Nuclear Safeguards Initiative on the June 1976 ballot and commit the state to an intensive program of energy development, guided by utility interests.

The second initiative provides that if the Nuclear Safeguards Initiative passes in June, but the Domestic Energy Initiative passes by a larger majority in November, the initial safeguards measure would be repealed. The Domestic Energy Initiative would then initiate the immediate construction of a statewide power system, with heavy reliance on nuclear energy. It would authorize the siting of offshore nuclear power plants, reactor fuel processing facilities, and waste storage facilities in California. Offshore supertanker terminals also would be built. The Domestic Energy Initiative would be administered by a committee of utility company executives and by the chairman of the state's Public Utilities Commission.

NUCLEAR LOBBY TACTICS

One well-documented and interesting example of how some professional proponents of nuclear power operate when they feel they are beyond public scrutiny is the case of the Atomic Industrial Forum's public affairs plan for 1975. Written by J. Lee Everett, III, chairman of the Forum's Public Affairs and Information Committee as a memorandum to his board of directors and adopted by the AIF, the document is a blueprint for an intensive public relations blitz. Although the memo itself acknowledges that "as a not-for-profit educational corporation the Forum is precluded from

---

*Among the largest contributors to the opponents war chests as of March 1976 were these donors: Bechtel Corp., $25,000; General Electric Corp., $20,000; Pacific Gas and Electric Co., $25,000; Southern California Edison, $50,000; and Westinghouse Electric Corp., $25,000.

lobbying," the tactics pursued by the AIF belie this. By an odd coincidence, in April 1975, just a few months after the memo reached the press, the AIF belatedly decided to change its registration from a non-profit organization to a trade association, which is allowed to lobby.

The AIF plans revealed that the company not only hoped to influence legislators in favor of the nuclear industry, but that it had devised a comprehensive strategy for managing and manipulating news media. This plan included the ghost-writing of articles for prominent people, the placement of articles in major publications including feature news syndicates, and the systematic distribution of industry information under the guise of news.

The AIF memo mapped detailed tactics in six major activity areas:

1. Nuclear Policy Information Service
2. Generation of Positive News Events
3. Contact with Other Influential Organizations
4. Field Service to Members
5. The Breeder Reactor
6. Speakers Bureau

Activities in area 1 were aimed at influencing "executive and legislative policy makers in Washington at both the staff and decision-making levels," by providing them information on the advantages of nuclear power. The program suggested establishing close personal contacts between AIF representatives and bureaucrats in the Nuclear Regulatory Commission, the Energy Resources and Development Administration, and members of crucial congressional committees and their staffs. A special mailing list was proposed to help direct pronuclear material to legislators, public utilities commission members, and local government councils. Along with the AIF's cornucopia of "important speeches, background papers . . . and special pamphlets concisely spelling out the economic and fuel supply benefits of nuclear power," the AIF would send suggested "rebuttals to major criticisms" of nuclear power. Apparently the AIF believed that the dissemination of a nuclear catechism, complete with orthodox correct responses, would help provide beleaguered nuclear technology with a common official line of defense. The AIF also intended to coordinate contacts by nuclear industry people and "independent experts" with governmental policy makers.

Another information service the AIF proposed was a state capital news network to "coordinate information to and from state legislative and executive bodies." One objective of the service would be to alert nuclear proponents throughout the U.S. of any state-level development likely to have significance in other states.

Under the rubric "Media relations," the AIF memo advocated increasing the Forum's widely distributed *Press INFO*, supplemented with "additional spot news stories, backgrounders and features." This program was necessary, according to the AIF, because, "The national media, with the middleman of the reporter and editor, cannot be relied on to publish a full and balanced account of nuclear power." Apparently, the AIF construes the journalist as an obstacle to be overcome in its management of the media. The AIF's aim seems to be a house-trained media that uncritically propagates the unadulterated industry version of nuclear power "realities."

Elsewhere in its memo the Forum states, "More work should also go into direct article placement, to minimize the *filtration factor* of the reporters and editors. Similarly, more direct use should be made of radio and television, rather than relying on their reporters for all news judgment" [emphasis added]. This allergy to reportorial interference with industry information is odd for a group ostensibly interested in distributing "balanced nuclear power information." One instance reported by James D. Sneddon in the *Beaver County* [*Pa.*] *Times* shows just how well this system can work. Sneddon relates that Frank Macomber of Copley News Service in the past simply rewrote news releases sent by the AIF to Copley, and that, Macomber in turn, is promoted by AIF. "In its latest *INFO*, it reports that 'reprints of a three-part series by Frank Macomber of the Copley News Service are available from *INFO*. The series looks at the country's commitment to nuclear power, the nuclear controversy and the information gap between the scientists and the public.' " Thus Macomber's work appears to be a series of articles by an independent source when, in fact, past stories have simply plugged the AIF position. The effect mushrooms when editorial writers take such stories and write opinion pieces favorable to nuclear power.

Under "Positive News Generation," the AIF memo recommends that the Forum "stage-manage news," by summoning the press to cover such news highlights as the press conference of a dozen key utility executives spirited to Capitol Hill by the AIF for visits to Congressmen. It also recommends using press junkets for top-level journalists, even though "the logistic and amenities expense can be fierce." This means that the AIF would probably furnish expensively catered food and drinks, hotel accommodations, and travel expenses for journalists' guided tours to various nuclear fuel cycle facilities.

Along with other AIF advice on news feature placement, the memo described a previous AIF media experiment in which the AIF hired "outside mat services" with which to "blanket small papers." Mat services are firms that presumably would turn AIF-written features into camera-ready synthetic news copy. The memo added, "The time is ripe for the PAIP [Public Affairs Information Program] staff to ghostwrite substantive features for well-known experts and to place them with major consumer publications. At the same time, we should aggressively seek space in these publications for unusual pro-nuclear articles."

The memo also recommended using the mass duplication of taped radio messages to transmit nuclear industry responses to critics or to distribute feature stories. "Surprising as it may be," said the AIF memo, "hundreds of radio stations will air taped messages coming from outside sources." These stations, perhaps unwittingly, would then be manipulated into executing the nuclear industry's public relations blitz.

The massive scale of the AIF's effort emerges in their plans for "Contact with Other Influential Organizations." The AIF proposed to contact directly and provide information on a *continuing basis* to a large selection of organizations important to the acceptance of nuclear power, including labor unions, segments of the financial community, professional organizations, and women's groups. Tactics suggested by the AIF include visits, briefings and seminars. "Specific goals," said the memo, "would be to place articles in *their* publications, arrange speakers for *their* conferences and generally encourage their vocal support of nuclear power." (Emphasis added.)

Among its "Field Services" to AIF members in the nuclear industry, the AIF plans to dispatch an elite corps of pronuclear public relations "troubleshooters" to allay public fears of nuclear power. These squads would be sent to help rescue "members with particularly pressing local problems — such as a serious moratorium threat . . . " Such PR agents are frequently much more intensively trained than the lay citizens they oppose. If successful, these outsiders might deny a local group their legitimate right to repudiate nuclear power.

Citing the far-reaching implications which "go or no go decisions to be made on the breeder" will have on nuclear power development, the Forum memo announced the initiation of a national breeder reactor information program that would serve as the primary channel of breeder information from industry to the media, government, and influential groups.*

Because successful nuclear advocacy often hinges on the availability of articulate speakers who can acquit themselves well in debate, the AIF intends to train a half-dozen people in "speaking and confrontation tactics" and to place a few experts on retainer so they can be sent to defend the cause on short notice. These speakers would also have to help the forum in other ways "such as writing letters to editors of key publications."

In general, the AIF memo revealed the clear intent to convert independent media into a tool for propagating nuclear industry views, and to convince public leaders to formulate policy on the basis of one-sided pronuclear information. Moreover, these are only the tactics that one nuclear lobby admitted to *in writing*. The methods they might actually employ in aggressively placing in-house articles in national publications can only be imagined.

Besides distributing unpaid nuclear advertising disguised as news, nuclear proponents have conducted huge formal advertising campaigns. Many of the ads have been placed by nuclear utilities, whose customers are not consulted about ad content although immense sums are spent on these controversial messages. Later, the utility claims a tax deduction for its "public service," and the

---

*The fact that the AIF feels it vital to coordinate breeder reactor information more than 20 years before the reactor is expected to be commercially available indicates what an integral role the breeder plays in nuclear industry plans. The prospect of high uranium fuel costs, uranium shortages, and no breeder reactor are powerful deterrents to potential reactor customers who are worried about the future availability of competitively priced nuclear fuel.

public pays in forgone tax revenue. In 1974, PG&E's nuclear advertising, including ads on the need for new construction programs, were budgeted at $400,000, excluding staff and administration costs.* Nuclear energy promoters have the resources to purchase costly television spots and buy two-page spreads in *Time* and *Newsweek*. Such an ad for nationwide distribution in *Time* can cost $70,000. Subsidized by public funds, utilities can also send free pronuclear material to the country's 64 million households. It would cost citizens millions to do a comparable mailing, excluding the substantial costs of production and labor.

Written in glib, folksy styles, many of these ads contain highly deceptive information and sometimes outright falsehoods on nuclear power. For example, Westinghouse Electric Corporation placed an ad in *Time* magazine (September 18, 1972) and in *Newsweek* (September 25, 1972) asserting that "the energy crisis can be averted by the breeder power plant." (See next page for text of ad.) The ad went on to say, "The 'energy crisis' means that the U.S. will not have enough usable coal, oil and natural gas left to keep going," and concluded by saying that breeder reactors will produce "essentially no radioactivity."

Each of these claims is incorrect. Even if breeder reactors are developed on schedule and become commercial in the 1990s, they would not be producing more than 2.5 percent of U.S. electrical generation — hardly a way to "avert" the energy crisis.[5] Solar energy by that time could be producing a ten times larger share.

The notion that the U.S. is about to exhaust its usable coal supplies is equally absurd. U.S. coal reserves are sufficient for hundreds of years, according to the U.S. Geological Survey. Even more dishonest was the Westinghouse assertion that breeders produce no radioactivity. Breeders produce *enormous* quantities of radioactivity, and their fuel is so radioactive that, unlike the current

*The company's overall advertising budget for 1974 was $1.9 million, according to a company official.[4] In a complaint lodged against PG&E on July 18, 1975, Project Survival, a citizens' group opposed to nuclear power, asked the state's Public Utilities Commission to order PG&E to stop promoting nuclear power at public expense. The complaint charges the company with improperly exercising its privileged monopoly position and power to influence media coverage of nuclear power. For its profile of PG&E's political advertising, and as a general reference on pronuclear advertising, the Project Survival brief is a valuable resource. The project's main office is at 366 California Avenue, Palo Alto, California.

TEXT OF WESTINGHOUSE BREEDER ADVERTISEMENT
[From Time magazine, Sept. 18, 1972]

[Text depicted on a Tombstone]
U.S. economy victim of energy crisis, Circa, 2000 AD.

FOR SALE: THE ANSWER — THE ENERGY CRISIS CAN BE AVERTED
BY THE BREEDER POWER PLANT: WESTINGHOUSE IS SPENDING
$60,000,000 OF ITS OWN MONEY TO HELP DEVELOP IT

The "energy crisis" means that the U.S. will not have enough
usable coal, oil and natural gas left to keep going. Something our
grandchildren would not thank us for.

Breeder power plants remove the threat

The "breeder" is a special kind of nuclear power plant. It's
strange: it makes more fuel than it uses.

(You don't refuel a breeder power plant when it's time to put
more fuel in; you refuel when it's time to take fuel *out.*)

A new natural resource — Our biggest

With breeder power plants, the U.S. will gain a new natural resource of energy for making electricity. A far greater resource than all
our coal, oil and gas combined. Enough, all by itself, to keep a
growing U.S. going for hundreds, probably thousands, of years.

That's why Westinghouse is spending the $60,000,000 of its own
money.

To help make the first U.S. breeder power plant a success.

P.S. — Less urgent than the energy crisis, but not to be overlooked: breeder power plants will produce no smoke; essentially no
radioactivity; much less "thermal pollution"; and electricity at lower
fuel cost. Prototype breeders are being built in France and England,
and one is operating in Russia.

generation of light water reactors, breeder fuel can reassemble itself
into a supercritical mass during a serious accident.

Defenders of Westinghouse may argue that the company really
meant breeders *release* essentially no radioactivity. This claim has
not been proven either, but if that is what was meant, it should
have been stated. Westinghouse certainly had the resources in its
reactor engineering division to check the technical accuracy of its
ads. One can only conclude this was another deliberate attempt
on the part of the nuclear advocates to mislead the uninformed
public about the dangers of reactors. Other misconceptions fostered

by the brief ad were that the breeder would produce less waste heat than fossil-fired plants. Actually, breeder reactors produce more waste heat than fossil plants.

Another particularly deceptive ad, titled, "Why Nuclear Power is the Solution to the Energy Problem," was widely distributed in Northern California by Pacific Gas & Electric Company. This ad is filled with sometimes subtle but crucial distortions of fact. Written in a disarmingly colloquial style, the ad intimates that without nuclear power, the customers risk brownouts and blackouts. A photo of an unprotected hand holding nuclear fuel pellets projects the notion that nuclear plants present no health hazards. And the complex issues of energy policy are reduced to simplistic slogans, as in the ad's first paragraph:

> The electric energy problem here in California is simply a matter of oil and natural gas shortages. The problem is going to plague us for some time to come, unless other forms of energy are used. The solution is to use energy wisely and to build more nuclear power plants. Nuclear plants are safe. They are practical. They are economical. They are environmentally clean. But they take time to build — about ten years.

So judging from this introduction, the case for nuclear power is an open-and-shut one. Yet every one of these assertions, save for the ten-year construction period, is wrong or highly controversial. Northern California has no electricity shortage, for example; it has excess generating capacity. And in the ad's first line, the company chooses to ignore that political decisions not to rigorously pursue energy conservation measures have as much to do with fuel shortages as do resource constraints.

Although the entire ad can be refuted virtually line by line, readers by now should have enough ammunition to do much of the job themselves.* Only a few passing comments will be made here. First, the assertion by PG&E that "the overwhelming majority of the scientific community" believes strongly in nuclear plants turns out to have been based on a survey of the American Nuclear Society membership. Next, the claim that "more than 95 percent of the original fuel is recycled for re-use" is blatantly untrue in that no reprocessing facilities are operating in the U. S.

---

*For a detailed rebuttal of the ad, see "How One Utility Hypes Nuclear Power," *Ramparts,* August 1974.

WHY NUCLEAR POWER IS THE SOLUTION TO THE ENERGY PROBLEM.

The electric energy problem here in California is simply a matter of oil and natural gas shortages. The problem is going to plague us for some time to come, unless other forms of energy are used. The solution is to use energy wisely and to build more nuclear power plants. Nuclear plants are safe. They are practical. They are economical. They are environmentally clean. But they take time to build—about ten years.

There's no mystery about nuclear power plants. There are 44 operating in the United States; more than that among other nations of the world. There are more than 100 nuclear-powered ships in the U.S. Navy; even more in other fleets. The nuclear industry has hundreds of reactor years of successful operating experience. The technology is proven.

Some people have questions about nuclear power. Some people give incorrect answers to those questions.

We at PG&E have had long experience with nuclear plants. We believe firmly in them. So do other utilities, world-wide, both government-owned and investor-owned. And so does the over-whelming majority of the scientific community.

Nine nuclear pellets produce as much energy as 30 barrels of oil.

NUCLEAR POWER AND THE FUEL SHORTAGE

At present most of PG&E's steam-electric power plants burn scarce and very expensive low-sulfur oil to generate electricity. We will have to buy about 20 million barrels this year and 35 million next year to meet our customers' electric energy needs. Our two-unit Diablo Canyon Nuclear Power Plant, now under construction in San Luis Obispo County, will displace a need for an additional 24 million barrels of oil every year in the future.

Delays in construction schedules of these and other nuclear units—delays, for a variety of reasons, over which utilities generally have little control—have had much to do with bringing about today's electric energy problems in California.

While nuclear power plants cannot solve the problem immediately, they can in time. As more come into service, they will free up large amounts of oil, significantly alleviating the aggravating long-range fuel shortage—gasoline and all.

NUCLEAR POWER AND SAFETY

The safety record of commercial nuclear power plants is unmatched in industrial history. Safety systems and their back-up systems function efficiently. There have been no nuclear-caused deaths. Not even a significant injury. (For comparison, about 54,000 Americans are killed every year in auto accidents; 3,000 die choking on food; 160 are killed by lightning.)

Actually, fissionable nuclear fuel for power plants is very dilute—so dilute that it's impossible to create an atomic explosion in a nuclear reactor.

With all the safeguards that are built into each nuclear power plant, the chance of a major accident is about one in a million.

NUCLEAR WASTE. WHAT HAPPENS TO IT?

When nuclear fuel is used, nuclear waste is created. But more than 95 per cent of the original fuel is recycled for re-use. The remaining waste is small—so small that such waste from a large nuclear unit operating for 30 years could be contained in a space no larger than a two-car garage. The waste is radioactive; but is treated as such. Very carefully. Safety first.

Used fuel is sealed in heavily-shielded, leak-tight casks and shipped to a facility which specializes in nuclear fuel reprocessing. Every safety precaution is taken to insure that no leakage occurs. Shipping and handling are carried out under strict regulations of the AEC and the U.S. Department of Transportation. After processing, the residual waste will be solidified and placed in secure, long-term storage under rigid government control.

NUCLEAR POWER PLANTS AND MARINE LIFE

Some people have voiced concern because some power plants discharge warm water back into natural water bodies. These power plants—

Figure 8-2. PG&E nuclear promotion advertisment

whether nuclear or fossil-fueled—use cooling water in steam condensers. In a nuclear plant the cooling water is only about 19° warmer when returned to its source, and otherwise is harmless. Where the water source is large enough and cold enough to receive and assimilate it, like the Pacific Ocean, it has no significant adverse effect on marine life. The only appreciable change is that in the immediate water discharge area the balance between warm water species and cold water species of marine life may shift in favor of those liking warmer water. In fact, after 24 years of scientific study and many more years of operating experience, it is clearly established that marine life near PG&E power plants tends to be *more* plentiful than it was originally.

NUCLEAR POWER-CLEAN, ECONOMICAL

For both environmental and economic reasons, nuclear power is the solution to the electrical energy problem.

Most hydroelectric power resources are already developed. Fossil-fueled steam electric plants consume scarce and increasingly costly oil and natural gas. Barring technological breakthroughs, geothermal energy can meet only a small part of future power needs.

Thousands of men work and live safely on nuclear-powered subs.

Fusion power is decades away. And other possible sources of energy, such as solar, tidal and wind power, are in experimental stages of development, and the latter two may never become practical for large-scale use. Coal can supply some help in California over the short run. But nuclear energy is the power source which has arrived.

Nuclear power is economical. For example, the electricity produced at PG&E's Nuclear Power Plant for $2.00 would cost $17.20 at a plant burning low-sulfur oil, at today's fuel prices.

Moreover, nuclear power generation is clean. Unlike burned fuels, it releases no combustion products into the environment.

NUCLEAR POWER AND INSURANCE

Some people say that private insurance companies won't cover a nuclear power plant. That's false. Private companies provide $110 million worth of liability insurance for each nuclear power reactor location. There have been no claims against nuclear power reactors. In fact, the insurance companies have been refunding part of the premiums paid by the utilities.

In addition, utilities pay the federal government for indemnity insurance coverage of $450 million for each reactor location. The federal indemnity program was created by Congress in 1957 (Price-Anderson Act) to help encourage development of a nuclear power industry in the U.S. It has been good business for the taxpayers. And it gives the public greater protection than separate homeowner insurance policies could provide. That's one of the reasons why your homeowner policies have a nuclear exclusion clause.

The government has collected millions in indemnity payments from utilities—about $90,000 a year per large reactor—and has never paid out one cent. No claim has ever been filed.

NUCLEAR POWER AND THE PUBLIC INTEREST

One of the big PG&E nuclear units at Diablo Canyon is planned for service next year, and the other unit in 1976. But it will take about ten years to build additional nuclear capacity—including the time it takes to find and acquire suitable sites and obtain clearances and approvals from more than 30 governmental and public agencies.

Every year of delay exposes all of us to shortages and higher rates, and further drains our diminishing fossil fuel resources.

The energy problem simply must be solved, and nuclear power will go a long way toward solving it. Electrical energy is essential to everybody, and especially to the young people who will be forming families and needing jobs. We don't intend to relax in our efforts to provide adequate and reliable service for all our customers in the future, just as we have provided it in the past. You can help now by conserving energy at home and on the job.

If you or anyone you know would like more information on nuclear power, PG&E will be pleased to provide it. Just write: PG&E Nuclear Information, 77 Beale Street, San Francisco, California 94106.

The assertion that "the technology is proven," is belied by the fact that emergency core cooling systems are still undergoing basic tests. Finally, readers knowing that the company's Diablo Canyon nuclear plants have sustained cost overruns of more than 100 percent should have no difficulty in evaluating PG&E's boast that nuclear plants are economical.

In addition to lavishly funded advertising campaigns and intensive nationwide public relations efforts, nuclear proponents have also suppressed information critical of nuclear power. Evidence of the AEC's efforts to squelch unfavorable safety test results and minimize the consequences of nuclear accidents have been amply documented in the *New York Times, Environment* magazine, and elsewhere.[6]

The record reveals a well-established policy of information manipulation, designed to keep from alarming the public with facts about reactor safety and environmental problems. The AEC obstructed the progress of public hearings on reactor safety; it censored AEC research reports; took reprisals against dissenters on its own staff; and terminated vital safety research that cast doubts upon fuel rod safety.[7,8,9] In addition, the AEC in 1965 conspired with the Atomic Industrial Forum in an effort to prevent the release of the Brookhaven National Laboratory's revision of its 1957 Brookhaven Report on the estimated consequences of a major reactor accident.[10] This successfully kept the report's awesome conclusions from the public for eight years, until the AEC was forced to divulge them in response to a Freedom of Information Act demand. The AEC also allowed substantial conflicts of interest to occur in its safety research program: General Electric and Westinghouse, both reactor manufacturers, were hired by the AEC to conduct critical tests of their own safety systems.[11] The questionable data produced in the process further obfuscated the true nature of reactor safety problems — from AEC regulators as well as from the public.

THE POWERS THAT BE

Like the AEC, utilities also have been known to engage in information suppression. Pacific Gas & Electric Co., for example, caused a critical documentary on nuclear power to be withdrawn from circulation and made unavailable for showing, even to some groups who specifically requested it. The documentary, "Powers That Be," was made by Emmy Award-winning producer Don Widener for KNBC-TV in Los Angeles, and was narrated by actor Jack Lemmon.

The film deals with the dangers of radiation, nuclear accidents, and waste storage along with the promises of clean power from several alternative energy sources. The documentary contained outspoken criticism of nuclear power, and was broadcast only once, on May 17, 1971, over KNBC-TV. It was never accorded a full national network showing by NBC.

During the film, Widener interviews James Carroll, a nuclear engineer at PG&E's Humboldt Bay nuclear power plant, and asks him about allegations of faulty fuel rods at the plant. Carroll, backed by PG&E, later complained that the film was irresponsible and fraught with half-truths. Worse, Carroll alleged, Widener had secretly tape-recorded a pre-interview conversation, and then had spliced the material into the film to make it look as though Carroll were evading the question on fuel rods. Commenting on Carroll's allegations, Elliot Kanter of *Ramparts* magazine wrote:

A careful viewing of "Powers That Be" shows that J. C. Carroll's words are perfectly synchronized with his lip movements — a virtually impossible triumph of editing if his answers had actually been taken from a surreptitiously taped conversation. Carroll, perhaps upset at reports of the bad impression he had made, delivered his protest without first having seen the film.[12]

PG&E sent complaints about Widener's film to prominent senators and to the Federal Communications Commission, which failed to sustain the company's contentions. The protests by PG&E charged Widener with bias, malice, and unprofessional conduct. He filed suit shortly thereafter against the company for $8 million. Widener's brief accuses PG&E of trying to discredit and defame him, thereby suppressing his film and damaging his career.

In 1972, PG&E filed a counter suit for $6 million, asserting that Widener doctored the film to discredit nuclear power, and demanding an injunction against distribution of the whole film. During the four years since its broadcast, it has been withheld from public exhibition by KNBC, and the only copies to circulate have been the few controlled by Jack Lemmon's Jalem Productions of Los Angeles. According to Jalem's contract with KNBC, even these could not be shown publicly until the one-minute Carroll footage was cut.

Although PG&E could have directed its legal assault on removing the disputed interview from the film, the company instead moved to block the distribution of the film in entirety, and while the case was in court, it unsuccessfully sought a "gag rule" to seal the court files and prevent Widener's attorney, David Pesonen, from discussing it with journalists.[13,14]

In November 1975, Widener recovered a unanimous jury verdict of seven and three quarters million dollars in San Francisco Superior Court. But in January, the trial judge reversed the jury and granted a judgement in favor of PG&E. Shortly before trial, PG&E had dismissed its cross complaint. Widener has appealed the judge's order, and the case is still pending.

The AEC, too, reportedly exerted pressusre to prevent the film's broadcast, and during the filming, summarily cancelled permission for Widener to interview several of its officials. By comparison, the AEC in fiscal 1973 had 120 films in circulation dealing with the peaceful uses of atomic energy, and the AEC estimated that 3.5 million people viewed these films in 1972, not including the *tens of millions* who must have seen them during their 155 television airings.[15]

Often it is not necessary for nuclear proponents to conduct elaborate public relations campaigns or go to the polls in order to exert powerful influences on public policy. In a sense, they simply whisper in policy makers' ears through the medium of the advisory committee, a venerable tool of public policy. More than 150 such committees exist on the national level dealing with energy policy. Although the 1972 Federal Advisory Committee Act requires that the committees represent labor, citizens' groups, and local government as well as industry, a study by Commissioner Ralls of the Michigan Public Service Commission found industry dominating the energy committees by a ratio of nearly fifteen to one. At taxpayers' expense, the industry boards were given nine times as much funding and eight times as much staff as the "consumer" committees.[16]

LEGAL CHALLENGES

The balance of political power that has existed for years between nuclear proponents and the nascent opposition movement has now begun shifting in favor of the nuclear critics, despite the heavy odds against which the critics have been fighting, and despite the temporary advantage bestowed on the nuclear industry by steep oil price increases. The most important basic cause of this power realignment is the growing awareness among citizens of nuclear power's dangers, and the alternatives to those dangers.* This citizen awareness has not gone unnoticed by legislators at every level of government, and several recently passed or proposed new laws reflect this situation and are likely to limit nuclear power development.

The Connecticut legislature, for example, adopted a bill in mid-1975 to set up a Nuclear Power Evaluation Council to determine the state's proper role in regulating nuclear facilities. The council is to be composed of experts in economics, environmental science, nuclear engineering, public safety, and radioactive material transport.

Connecticut also passed a law in September 1975 prohibiting political, promotional, or "good will" advertising at customer expense. From now on, the cost of such ads will have to be borne by utility stockholders. (Ads are considered political if they try to sway public opinion on controversial issues; promotional ads are those geared to increase power consumption.)

Other significant new legislation in 1975 was enacted in Vermont and Wisconsin. The Vermont legislature voted itself the right to veto new nuclear projects, and the Wisconsin Public Service Commission on May 1, 1975 asserted that it has the right to review nuclear power plant safety considerations because of their possible impact on electricity rates. The commission used this apparently circuitous legal route to claim safety review authority because federal courts have insisted in some instances that the formulation of rules and regulations on nuclear safety is the exclusive province of the federal government.

---

*Citizen opposition to nuclear power is discussed separately in Chapter 13.

LEGISLATIVE CHALLENGES

Among legislative proposals currently pending on the national level, the Nuclear Energy Reappraisal Act, introduced in the Senate by Senator Mike Gravel (S. 1826), and in the House by Congressman Hamilton Fish et. al. (H.R. 4971) is certainly one of the most important.* It would stop the growth of nuclear fission power in the U.S., pending the outcome of a comprehensive five-year study of the nuclear fuel cycle. Although passage is far from imminent, the bill is extremely useful in that it forcefully delineates major areas of concern about nuclear power and focuses attention on specific issues whose resolution ought to be a prerequisite for fission plant operations.

The act would authorize $75,000,000 over five years for an independent study of nuclear fission power by the U.S. Office of Technology Assessment. The licensing of new nuclear plants would be banned for five years and would not be allowed to resume unless, after the study, Congress determined that detailed provisions on power plant safety systems, radioactive waste management procedures, and plant security systems have been met. In order for licensing to resume, Congress would also have to conclude that "nuclear fission plants are clearly superior to other energy sources, including renewable energy sources." Five years after passage of the act, in the absence of congressional approval to license new plants, existing plants would be derated and ultimately phased out of operation.

The five-year mandated study would consist of a comprehensive investigation of the entire uranium fuel cycle from mining through fuel reprocessing and waste management. It would utilize testimony taken from independent scientists, engineers, consumer, and environmental representatives, but the Office of Technology Assessment would be barred from entering into contracts with, or relying primarily on, "the expertise of any industry or company which provides materials, management capabilities, research, or consultant services for nuclear fission power plants or which otherwise might have an interest in perpetuating the nuclear industry."

The office would prepare annual reports in addition to its

---

*The complete text of the Act will be found in the appendix.

final report, and would provide the opportunity for annual public hearings on the progress of the final study, as well as for informal public hearings throughout the study process. The annual reports would have to include any nuclear fuel-cycle safety information which previously had not been publicly known.

The office's final report would contain a recommendation to Congress on whether or not to approve the licensing of new nuclear plants. The recommendation would be based on consideration of major health issues such as "the short-term and long-term genetic effects of low-level radiation," and major unresolved economic issues. The latter include (1) the long-term costs and availability of uranium fuel; (2) "the cost implications of the frequent shut-downs of nuclear plants, including the costs of shut-down and start-up, inspections, the cost to consumers of purchase of alternative power during shut-down, unemployment benefits and other costs of unemployment that result from shut-downs"; (3) the probable costs and wisdom of relying on nuclear power, given that serious safety or sabotage problems at one plant could require the shut-down of all similar plants; (4) the costs of safeguards; (5) the costs of a nuclear transportation system; and (6) the costs of decommissioning existing fission power plants.

The office would also be charged with assessing the risks of nuclear proliferation, and the financial-technical capabilities of utilities, "as institutions" to operate nuclear plants safely "in light of the high costs of safety measures and the 861 AEC-documented abnormal events in utility operated nuclear plants in 1973." An additional assessment would be required of the reactor licensing process "which have permitted nuclear plants to be built over geologic faults and in other unsafe locations. . . . "

Another legislative effort to control the spread of nuclear power is the proposed Plutonium Recovery Control Act of 1975, introduced by Representative Les Aspin et al. (H.R. 3618, 4945, 4946, 6394). This bill would ban the licensing of certain activities relating to plutonium until explicitly authorized by Congress and would prohibit the licensing of plutonium–fueled reactors until a three-year study could be completed by the Office of Technology Assessment.

Nuclear proponents also have introduced proposed legislation on nuclear power, and a number of these bills are aimed at expediting the licensing of nuclear power plants. In this category are the Ford Administration's Energy Independence Act of 1975. This bill would allow early power plant site work in some circumstances and would alter regulatory practices governing electric utilities. Another bill, by nuclear proponent Representative Mike McCormack, the Nuclear Power Plant Siting Act (H.R. 3734), would authorize the Nuclear Regulatory Commission to issue licenses to plants conforming to state or regional licensing criteria.

A third bill, the Nuclear Facility Licensing Act (H.R. 3995), would permit the issuance of combined construction and operating permits for certain plants, instead of requiring separate permits and review processes. The bill would also end mandatory hearings now required prior to the issuance of construction permits, and would eliminate review by the Advisory Committee on Reactor Safeguards. Needless to say, environmental groups will strongly oppose such efforts to speed up nuclear fission plant construction.

CONCLUSION

The Atomic Industrial Complex is wielding its well-integrated economic and political power to insure that the U.S. makes a large commitment to nuclear power. The nuclear critics have replied via the initiative process, which enables them to take their case directly to the people instead of submitting it to legislatures that are subjected to extensive nuclear lobby pressure. Initiative efforts are under way in Washington and Oregon and environmentalists are planning initiative campaigns in fourteen other states. In addition, more traditional legislative sallies are being launched by both sides as the nuclear controversy heats up.

To date, the nuclear lobbies have faithfully represented the huge banks and other members of the energy cartel, helping them insure that vast capital spending on nuclear power continues. Working through a large array of powerful industry front organizations, the lobbies' atomic power plays have tried to confuse the public about nuclear power's costs and danger, and have also dealt setbacks to cheaper, more valuable alternatives. These tactics harmonize well with the energy interests' goal of keeping energy prices high and overcharging the public for this essential commodity.

NOTES

1. *The Energy Directory* (New York Environmental Information Ctr., Inc., 1974).

2. Chauncey Starr, "A Strategy for National Electricity Production," presented to the Joint Committee on Atomic Energy, Washington, D.C., July 10, 1975.

3. *A Summary of Program Emphasis for 1975*, Electric Power Research Institute, 3412 Hillview Ave., Palo Alto, California 94303.

4. "PG&E Attacked on Advertising," *San Francisco Chronicle*, Feb. 19, 1975.

5. Starr, *op. cit.*

6. David Burnham, "AEC Files Show Effort to Conceal Safety Perils," *New York Times*, November 10, 1974.

7. Daniel F. Ford and Henry W. Kendall, "Nuclear Misinformation," *Environment*, Vol. 17, No. 5, p. 17.

8. Joel Primack, "The Nuclear Safety Controversy," *Chemical Engineering Progress*, Vol. 70, No. 11 (November, 1974).

9. Robert Gillette, *Nuclear Safety* series, *Science*, No. 4051-4054, September, 1972.

10. Burnham, *op. cit.*

11. Gillette, *op. cit.*, No. 4054.

12. Elliot Kanter, "Powers That Be: The NBC Documentary You Never Got To See," *Ramparts*, August 1974.

13. "PG&E Move for Gag Rule in Suit," *San Francisco Chronicle*, Aug. 2, 1974.

14. "PG&E Loses Move on Suit," *San Francisco Chronicle*, August 8, 1974.

15. *Congressional Record*, June 22, 1973, pp. E 4306-7.

16. "Who Has Uncle Sam's Ear on Energy?" *People & Energy*, July, 1975, p. 4.

NUCLEAR POWER REACTORS IN THE UNITED STATES

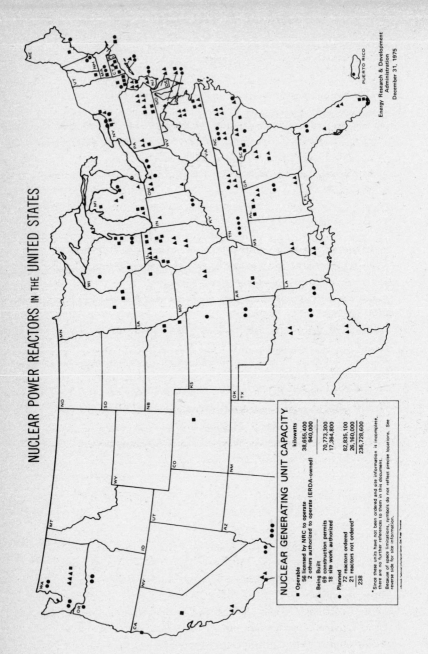

NUCLEAR GENERATING UNIT CAPACITY

kilowatts

■ Operable
56 licensed by NRC to operate 38,655,400
2 others authorized to operate (ERDA-owned) 940,000

▲ Being Built
69 construction permits 70,273,300
18 site work authorized 17,364,800

● Planned
72 reactors ordered 82,835,100
21 reactors not ordered* 26,160,000
238 236,728,600

*Since these units have not been ordered and site information is incomplete,
there are no further references to them in this document.

Because of space limitations, symbols do not reflect precise locations. See
reverse side for site information.

Energy Research & Development
Administration
December 31, 1975

Map 9-1. Nuclear power reactors in the U.S.

9

NUCLEAR INSECURITY

Imagine the morning after a nuclear explosive has destroyed half an American city. How are we going to [respond] if we cannot tell whose nuclear explosive it was? Or even if we could tell, but it turned out to be an organization, such as might exist in the future [sic] — an organization with dedicated people but no clearly defined national territory.

— Fred C. Ikle, Director U.S. Arms
Control and Disarmament Agency
New York Times, May 11, 1975

. . . political outlaws armed with stolen nuclear waste materials, let alone nuclear weapons, could hold entire cities hostage by threatening to pollute the area with atomic wastes unless their political demands were met.

— James Reston
New York Times, May 11, 1975

*   *   *

To enter a modern nuclear plant, such as the Rancho Seco plant near Sacramento, California, one needs to pass two barriers and a guard station equipped with an electronic warning system and video screen. Armed guards patrol the fenced perimeter. Coils of thick barbed wire rim the chain-linked fence. On the way out of the plant, personnel and visitors must pass a metal detector that bleats at keys and coins in their pockets. This is to make sure they're not making off with any radioactive materials that could be turned into dangerous weapons, and that they didn't inadvertently pick up any radioactive contamination. Would all those measures work to stop a determined band of armed attackers, bent on seizing the plant or making off with some of its radioactive material?

*   *   *

The American people — many still misled by the false promises of cheap nuclear power — now are being asked by their government and by utilities to accept an era of nuclear insecurity, filled with risks about which we have never been adequately informed. Few of us today *really* understand the dangers of nuclear sabotage, nuclear terrorism, nuclear transportation accidents, or nuclear proliferation, nor how these add to the already frightening risks discussed in previous chapters. Yet the making of an illicit nuclear bomb, the intentional destruction of a nuclear facility, or the continuing spread of nuclear material abroad, *each* could have consequences far greater than a nuclear reactor meltdown, and none of the three developments mentioned is a "far fetched" eventuality.

THE BASIC PROBLEM

Explosive plutonium and highly enriched uranium-235 are in general circulation in the U.S. and throughout the world as a result of the routine operation of the nuclear industry. These materials are bomb-grade, which means they can be used to make nuclear explosives, either directly or after chemical processing. Hundreds of pounds of enriched uranium have — by the AEC's own admission — vanished in recent years, as have smaller quantities of plutonium.

Plutonium not only can be made into bombs, it can also be used directly as a radiological contamination weapon of mass destruction. As an oxide or nitrate, plutonium could be scattered manually. A moving vehicle or a tall building could help make effective plutonium dispersal easy. Once suspended in the air in finely divided particles, a small amount of plutonium poses a grave threat to anyone who breathes. Several pounds of plutonium for use as a weapon could be easily carried and concealed by a person without special shielding, because the alpha ray emissions of plutonium are lethal only when particles are inhaled or otherwise actually enter the body. Alternatively, using widely available and accurate technical information on bomb construction, a nuclear thief could use plutonium or highly enriched uranium to build an atomic bomb of devastating power. Possession of the bomb would give its controllers the power to threaten governments and whole cities.

Before it was abolished, the AEC was severely criticized for the laxity of its procedures for insuring the protection of weapons-grade "special nuclear materials" (SNM). In response to the criticism, the AEC commissioned a report entitled, A Special Safeguards Study, by Dr. David Rosenbaum, a former FBI official, and several other specialists.[1] The Rosenbaum study concluded, "The potential harm to the public from the explosion of an illicitly made nuclear weapon is greater than that from any plausible power plant accident, including one which involves a core meltdown and subsequent breach of containment."

The potentially catastrophic consequences of nuclear insecurity have been aptly described in more detail by two experts on the safeguarding of nuclear materials whose criticisms of safeguards helped lead to the Rosenbaum study. University of Virginia law professor Mason Willrich and nuclear physicist Theodore B. Taylor stated in *Nuclear Theft: Risks and Safeguards:*

> A nuclear explosion with a yield of one kiloton could destroy a major industrial installation or several large office buildings costing hundreds of millions to billions of dollars. The hundreds or thousands of people whose health might be severely damaged by dispersal of plutonium, or the tens of thousands of people who might be killed by a low-yield nuclear explosion in a densely populated area represent incalculable but immense costs to society.[2]

And that is the assessment of two men who, nonetheless, favor an increasing domestic reliance on nuclear power because they feel that nuclear safeguards can be made sufficiently fail-safe. Their conclusion is hard to accept, given the risks of nuclear insecurity.

The raw materials for nuclear weapons are abundant and vulnerable to theft as a result of current nuclear reactor operations in the U.S. Almost all U.S. reactors are fueled by low-enriched (2 to 4 percent) uranium and produce plutonium in their fuel rods during power production. Thieves are not likely to be tempted by the low-enriched fuel or by these plutonium-bearing rods. The rods are poisoned by gamma-ray emitters like cesium-137, as well as by a large assortment of other toxic radioisotopes. Consequently, the rods can only be handled by robotized equipment in the heavily shielded vaults of a nuclear fuel reprocessing plant. But highly enriched uranium, and plutonium after extraction from spent fuel rods, can be readily used in nuclear weapons.

In addition to ordinary reactors, fueled by low-enriched uranium, at least one high temperature gas reactor (HTGR), fueled by highly enriched uranium-235 and thorium, is being completed. The risks of a plutonium or uranium theft will be much greater if breeders and high temperature gas reactors eventually become prevalent, as many nuclear proponents expect.

The breeder reactor, not yet commercially developed, would take a fuel load of at least twenty-five hundred kilograms of plutonium and five to ten times that much uranium oxide (in the form of natural uranium or depleted uranium enrichment tailings); its initial fuel load also may be enriched uranium. An HTGR may require a fuel load of a thousand kilograms of highly enriched uranium. This enriched uranium would be difficult for thieves to use once it was converted to HTGR fuel granules because the granules are coated with an extremely durable carbide shell. Nevertheless, widespread reliance on HTGRs would insure that a great deal of enriched uranium was flowing through early stages of the nuclear fuel cycle at all times. If obtained in the early stages of the HTGR fuel cycle, highly enriched uranium could be used directly for bombs with no further processing.

Copious amounts of such bomb-grade material are currently produced in the United States. The Energy Research and Development Administration's gaseous enrichment plant at Portsmouth, Ohio, produces enriched uranium as a uranium-fluoride solid. This uranium fluoride can be used to make fuel for university research reactors in the U.S. or abroad; for fuel in nuclear submarines; or for making HTGR fuel. If intercepted, uranium fluoride, after simple chemical processing, can be used for bombs. If thieves wanted to avoid the bother of processing, they could steal the uranium after it had been treated at a conversion plant in Apollo, Pennsylvania; Erwin, Tennessee; or Hematite, Missouri. Conversion plants process the uranium fluoride to make uranium oxide powder or uranium metal that can be used directly in bomb-making.

In addition to the material leaving enrichment plants, fuel fabrication plants (located in Crescent, Oklahoma; Lynchburg, Virginia; New Haven, Connecticut; and San Diego, California) handle thousands of pounds of enriched uranium — enough for hundreds of bombs. Large quantities of enriched uranium can also be found at half a dozen fuel scrap recovery plants throughout the country.

If commercial fuel reprocessing facilities become operable, they, too, will produce highly enriched uranium, as well as plutonium. And with recycling, the plutonium would routinely travel from storage to fuel fabrication plant, and from reprocessing plants to storage or fuel fabrication facilities. Thefts could be attempted at storage areas, at reprocessing plants, or while material was in transit. Even though no commercial plutonium reprocessing is going on now, some recovered plutonium already is in storage.

NUCLEAR THEFT

Losses of nuclear materials inevitably occur during processing, because input and output quantities have to be estimated, and because the chemical processes used are not perfect. All plant materials accounting systems routinely allow for small margins of error in processing nuclear materials. This situation could be exploited by an adroit thief. If thievery is conducted in small quantities over a long period of time, the thefts might be impossible to detect.

The risks of theft are sure to increase greatly as the nuclear industry expands. According to AEC projections, more plutonium will be in the possession of private industry after 1980 than in the U.S. government's entire nuclear weapons program.[3] With a hundred nuclear fission plants in operation in the U.S. by 1980, fifty thousand pounds of plutonium will be produced annually, if current capacity factors obtain. Researchers of the Ford Foundation's Energy Policy Project estimate that two hundred thousand pounds of plutonium will be reprocessed by the mid-1980s if plutonium recycling is approved.[4] By the year 2000, as much as 2.2 million pounds of plutonium would be coursing through the arteries of international commerce, while according to AEC projections a thousand large nuclear power plants would be operating in the U.S.

As many as fifteen uranium enrichment plants, twenty fuel fabrication plants, and twenty fuel reprocessing plants would be needed to supply these plants and the reactors in foreign countries that would be depending on the U.S. for their nuclear material needs.[5] Hundreds of air, rail, or truck shipments of plutonium would be criss-crossing the U.S. by then, an open invitation to sabotage, accidents, and diversion.

Nuclear proponents believe that by the time massive expansion of the nuclear power industry occurs, nuclear safeguards able to contend with the added risks will have evolved. But most nuclear critics question that optimism. Military weapons, for which high security is provided, have not been well protected against theft.

In August 1974, ninety antitank missiles were reported missing from the U.S. Army Depot at Miesau, West Germany.[6] Even nuclear weapons storage sites in Europe have been judged vulnerable to guerrillas.[7] If the U.S. military has difficulty in assuring the safety of its ordnance from diversion, how is the civilian nuclear industry likely to provide adequate protection for its weapons-grade materials? This may be one of the reasons why the Ford Foundation Energy Policy report concluded, " . . . even a super-security system cannot prevent nuclear theft by highly determined and clever thieves." Today, however, the U.S. is far, far from having a "supersecurity system."

ACCOUNTING SAFEGUARDS

Complex accounting procedures have been set up by the AEC and by ERDA to detect the loss, theft, or pilferage of nuclear materials. Yet, as of 1971, the AEC was unable to account for 1 percent of the plutonium it handled.[8] If a million kilograms of plutonium are processed by the year 2000, with losses on a 1 percent scale, enough plutonium could be unaccounted for to build two thousand atomic bombs. But even the 1 percent loss-level has been disputed: According to the former director of the AEC's Office of Safeguards Materials Management, "We have a long way to go to get into that happy land where one can measure scrap effluents, products, inputs and discards to a 1-percent accuracy."[9]

Already, tons of radioactive materials have been unaccounted for at enrichment plants.[10] Although most of that material probably was not weapons-grade, losses of this size indicate that the nuclear industry is far from achieving adequate controls over its nuclear materials. On one occasion, two hundred twenty pounds of enriched uranium was reported missing and unaccounted for from a fuel fabrication plant.[11] Former AEC national security chief E. B. Giller told a Senate government operations subcommittee, "the AEC cannot be sure that two hundred twenty pounds of uranium unaccounted for over five years at a Pennsylvania facility was not stolen." He cited experts who concluded *it is likely* (but not certain) that the shortage resulted from the manufacturing process or from errors in measurements.

Losses of this size evidently troubled AEC officials and prompted some unnecessary "Cold War-ish" suspicions. When one hundred thirty-two pounds of uranium-235 was discovered missing from a fuel fabrication plant in Apollo, Pennsylvania (presumably the same plant as in the previous incident), the AEC for a while suspected that the material had been diverted to China for use in a nuclear explosive and so the agency spent five million dollars searching for the uranium. Later, on a high altitude overflight by a reconnaissance aircraft, the U.S. discovered China had built its own enrichment plant and would not need to steal U-235. In another case, the *New York Times* reported that there have been times when the Kerr-McGee fuel fabrication facility in Oklahoma was unable to account for up to sixty pounds of plutonium.[12]

CURRENT SAFEGUARDS

Three sets of Nuclear Regulatory Commission regulations summarize most of the current legal framework for dealing with nuclear insecurity. These regulations, *Licensing of Production and Utilization Facilities, Special Nuclear Materials,* and *Physical Protection of Plants and Materials,* can be found in Title 10 of the U.S. Code of Federal Regulations-Energy, Parts 50, 70, and 73.[13] Among its many detailed requirements governing the acquisition and handling of special nuclear materials, Part 70 requires licensees to "establish materials accountancy allowing the loss of no more than

Table 9-1   Transportation of Radioactive Materials in the Nuclear Fuel Cycle

Radioactive Material	Point to Point Movements	Mode of Transport	Required Packaging
Uranium or thorium ore	Mine to mill	Open truck	None
Ore concentrates (Uranium oxide or thorium oxalate)	Mill to refining and conversion plant	Truck	Drums
Natural UF_6	Refining and conversion plant to enrichment plant	Truck	Pressurized cylinders
Enriched UF_6 (recycle)	Reprocessing plant to enrichment plant	Truck	Pressurized cylinders
Enriched UF_6	Enrichment plant to materials processing plant	Truck	Pressurized cylinders and protective packaging
Thorium nitrate[a]	Refining plant to materials processing plant	Truck	Drums
Uranium (233[a] or 234) or plutonium oxides[b] (fissile)	Materials processing plant (or fuel reprocessing plant) to fuel fabrication plant (may be at same site)	Truck (if offsite)	Steel pails or sealed metal cans within gasketed steel containers supported inside steel drums
Thorium oxide[a] (fertile)	Materials processing plant to fuel fabrication plant	Truck	Drums
Transuranic wastes	Fuel fabrication plant (or fuel reprocessing plant) Federal repository	Rail[c] or truck	Drums and protective packaging
Irradiated fuel	Nuclear power reactor to fuel reprocessing plant	Rail[c] or truck	Specially designed and approved shielded casks
Recovered thorium[a]	Fuel reprocessing plant to	a	a
High-level wastes	Fuel reprocessing plant to Federal repository[d]	Rail[c] or truck	Specially designed and approved shielded casks
Beta-gamma wastes	From essentially all parts of the fuel cycle to commercial burial grounds	Truck	Drums and protective packaging

a To the present there has been limited experience in shipping uranium-233 and thorium compounds; this will become important when a large number of HTGRs are operational.

b To the present there has been limited experience in shipping plutonium compounds; this will become important plutonium recycle in LWRs wide use.

c Trucking to railhead may be necessary.

d The Federal repository is expected to be operational in the early to mid-1980's; in the interim, high-level fuel reprocessing wastes are being stored onsite.

e Currently shipped to burial grounds.

(1) 200 grams of plutonium or uranium-233, (2) 300 grams of highly enriched uranium-235 or (3) 9,000 grams of low-enriched uranium-235." But these apparently strict rules have a built-in loophole. Operators of nuclear facilities can have the loss-limits waived if they declare themselves unable to meet the limits and promise to improve their material control systems to meet those requirements.

The same sort of loophole applies to the limits of error permitted to nuclear facility operators in measuring their radioactive materials. For example, although the rules state that records on unaccounted-for plutonium, uranium-233, and highly enriched uranium-235 must be accurate to within half a percent, this requirement, too, can be waived. Federal regulations also exempt facilities handling small quantities of uranium and plutonium from stringent control requirements. Yet what the Nuclear Regulatory Commission regards as small quantities of nuclear materials still are potentially lethal and hazardous. A licensee handling less than five kilograms of 20 percent enriched uranium and less than two kilograms of plutonium would be exempt. Yet a potent indoor contamination weapon could be made with a gram (.03 ounce) of plutonium or less.

The physical security regulations set requirements for (1) the shipment of nuclear materials, (2) the physical protection of nuclear plants and materials, and (3) the record keeping on nuclear materials. These regulations reflect the fact that the transportation of nuclear materials is a weak link in the safeguards chain.

Truck shipments of nuclear materials must either have an escort with at least two armed guards, or be carried in a specially designed truck or trailer that can be immobilized in an emergency. Both the shipping vehicle and the escort must have radiotelephones. Obviously, none of these measures would neutralize determined attackers. A squad of guerrillas could overcome the two guards before reinforcements arrived. After blasting open the shipping vehicle, the guerrillas could transfer their booty to a getaway truck.

According to Nuclear Regulatory Commission rules, licensees merely have to protect their facilities against sabotage and thefts by using a security organization acceptable to the NRC. Although since June 1974 nuclear utilities have been required to arm their guards, these guard forces are so small they could easily be overpowered. Yet when the utilities were ordered to take even this elementary precaution, "high nuclear industry sources" cited by the *New York Times* reportedly objected that the order might cause new dangers "because the calibre of industrial guards is not very high — they are not exactly an elite corps."[14] Officials of Con Edison of New York also complained before complying with the order because of the danger that gunfire might damage power plant facilities. (It clearly could do so.)

These are some of the reasons why pro-nuclear safeguards experts Willrich and Taylor concede that, "Taken together, present U.S. safeguards do not constitute a *system*. An effective system of safeguards may evolve, if present trends continue."

Present safeguards are so deeply flawed that even if they were rigorously introduced throughout the world they would themselves create new risks on a grand scale. Safeguards depend on the implementation of protective measures by reliable, trusted people. But even the U.S. nuclear weapons program, which employs intensive background checks prior to issuing security clearances, has found it necessary to weed out about 3,600 undesirable employees every year because of newly discovered mental illness, alcoholism, drug abuse, and other reasons.[15] Thus it appears that imperfections in the state-of-the-art of security checking might easily allow a nuclear facility or transport security force to be infiltrated by an organization bent on nuclear theft or sabotage.

CIVIL LIBERTIES

Paradoxically, the more safeguard measures proliferate, the more they will inexorably threaten civil liberties. A federal nuclear police force has already been proposed and other measures are likely, such as the use of expanded security checks; wiretapping and other covert surveillance; the increased use of informers and infiltration and surveillance to check on people deemed suspicious; and more dossiers on "dissidents."

If nuclear safeguards were breached, this could readily be used as a rationale for the declaration of martial law and the suspension of civil rights until the threat had been neutralized. Mention of this possibility and the proposed repeal of laws governing illegal searches has been made in the AEC's draft environmental impact statement on plutonium recycling.

If a critical mass of bomb-grade uranium or plutonium were known to have fallen into the hands of a malefactor, a martial law declaration could easily be justified to expedite searches, facilitate evacuations, and maintain general order among the frightened populace. But the state of emergency could lead to forcible evacuations of cities or to large-scale occupations of areas by federal troops with orders to shoot first and ask questions later.

Martial law would also make possible preemptive arrests, coercive interrogations including torture, and the detention of suspects. In situations of martial law, the military has the power to arrest, confine, and try suspects by military commissions which are beyond normal judicial control and are not even subject to conventional military law. There would be no impartial panel of peers to rule on suspects' guilt or innocence, and no statutory requirement for a unanimous decision on that question. The emergency would also raise the spectre of citywide search-and-seizure missions by military and police, without regard for traditional Fourth Amendment protection against warrantless searches.

Writing in the *Harvard Civil Rights-Civil Liberties Law Review,* Russell W. Ayers states, "Once a quantity of plutonium had been stolen, the case for literally turning the country upside down to get it back would be overwhelming."[16] Therefore the theft or potential theft of plutonium could not only threaten public order, lives, and safety, but America's social and political institutions, too.

Safeguards will also entail large monetary costs which cannot yet even be accurately estimated. Willrich and Taylor have suggested that adequate safeguards could be created for a small percentage of the overall cost of nuclear power. But the dangers that safeguards must counter are multiplying rapidly. Ever greater volumes of special nuclear materials are being produced and handled daily. Breeder and high temperature gas reactors loom on the nuclear horizon. Knowledge about special nuclear material is spreading throughout the world, and guerrilla organizations are becoming more formidable.

Therefore, the cost of adequate safeguards will soar. Willrich and Taylor have estimated that $79 million annually would be required in 1980 — about 1 percent of the expected value of nuclear electricity generated annually then — just to fund a nuclear security force in the U.S. The security force would include a hundred full-time security employees (i.e., about thirty per eight-hour shift) for each major nuclear facility, such as a reprocessing or fuel fabrication plant. But many other unknown expenses will push the total costs for safeguards even higher.

While protesting basic safeguards as unnecessary, nuclear industry officials contend that the protection of nuclear facilities is a national defense function. Consequently, they assert, the ordinary taxpayers should absorb the costs of safeguards. Needless to say, trying to safeguard the U.S. against the existing risks of international hostilities is far different from defending it against the arbitrarily imposed risks created by the nuclear industry's pursuit of maximum profits. Those totally unnecessary risks could be avoided by using other energy technologies.

The nuclear proponents' strategy is to try to justify the proposed security subsidy by suggesting, "Since we're all consumers of electricity, we all should pay the cost of protecting the means by which it is generated." The catch is that we're not all consumers of *nuclear* electricity, because nuclear sources produce only a small fraction of the nation's electricity.* And if we properly internalize *all* the costs this industry is generating back within the industry itself, the "attraction" of nuclear fission would be nil.

BOMBS AND BOMB-BUILDING

Knowing that special nuclear materials (SNM) are abundant, imperfectly protected, and suitable for atomic weapons, it is especially unsettling to know how small a quantity of SNM must actually be stolen for the preparation of a potent nuclear weapon.

For a nuclear weapon to explode, an atomic chain reaction, or series of collisions, must occur. (See Chapter 2.) When a stray neutron hits a plutonium-239 atom, for example, and causes it to split, other neutrons shoot out of the plutonium atom's nucleus along with tremendous amounts of energy. If the mass of plutonium is large enough and dense enough, the newly freed neutrons from the first fission will collide with other plutonium atoms in a self-

---

The nuclear power industry has yet to produce a *single* net kilowatt of electricity, once the cumulative energy inputs expended for uranium enrichment and other aspects of the nuclear fuel cycle have been subtracted.

sustaining chain reaction. For different kinds of fissile materials, there are different minimum amounts or "critical masses" that must be present to insure that enough free neutrons are available to produce fast fission. If the fission occurs too slowly, the plutonium mass would be blown apart by the very onset of fission in a "fizzle yield," before much of the plutonium's explosive potential had been released.

In addition to its dependency on the mass and density of fissionable material, the critical mass of a radioactive substance depends on the ability of the metallic shell or "tamper" holding the bomb to reflect neutrons during the chain reaction. The greater the reflecting ability, the more intense the chain reaction. Therefore, the amount of fissile material needed for a bomb depends on absolute physical limits and on the designer's skill. Although it is theoretically possible to have a critical mass with nuclear material the size of a pea, larger amounts are needed in practice. For one of the simpler atomic bombs, the critical masses inside a thick beryllium shell are four kilograms of plutonium, or eleven kilograms of uranium-235, or four and a half kilograms of uranium-233.

Compression af fissile material by a surrounding sphere of explosives, so as to increase the density of the fissile material, decreases the minimum amount of special nuclear materials needed for a critical mass. Trigger quantities of plutonium for one of these "implosion devices" might be as little as two kilograms, if the bomb were made by a sophisticated weapons expert. Likewise, five kilograms of highly enriched uranium-235 might be enough for a bomb.

Although more material is needed to make a uranium bomb than a plutonium bomb under otherwise equal conditions, the uranium-bomb-fabrication process would be much less hazardous to the bomb-maker. Whereas plutonium would have to be handled within a glove box, subcritical amounts of uranium-235 could be handled long enough to make a bomb without special shielding.

SKILLS REQUIRED FOR BOMB-MAKING

Recently, when a twenty-year-old college chemistry student was challenged to design a workable atomic bomb within five weeks, he succeeded in producing a design for a bomb capable of exploding with a force equal to millions of pounds of TNT.[17] Although the execution of the design would require daring, skill and effort, Willrich and Taylor find,"It is difficult to imagine that a determined terrorist group could not acquire a nuclear weapon manufacturing capability once it had the required nuclear weapon materials."[18] Accurate information on how to do it is widely available in unclassified literature.

Whereas today, guerrillas are unlikely to steal the slightly enriched uranium fuel used in most nuclear plants because there would be no way for them to enrich it, in the future, the availability of new laser- and gas-centrifuge enrichment techniques may turn low-enriched or even natural uranium into attractive targets for guerrillas and thieves.

Already, the threat of nuclear theft and terrorism is growing as guerrilla groups become more sophisticated and numerous. Yet we are only in the first phase of exposure to nuclear terror. In April 1974, in what may be the first use of a radiological weapon, two trains in Austria were found contaminated by radioactive iodine-131 after anonymous warnings were issued by a man calling himself a "justice guerrilla."[19]

The gallery of potential thieves who might be tempted by nuclear materials is a large one. The roster includes psychotics, disgruntled employees, blackmailers, guerrillas, agents of non-nuclear nations, a rebellious clique within a government wishing to implement a coup, even dishonest managers of a nuclear facility wishing to sell their loot on the black market.

Considering the threat from guerrillas alone, we can easily imagine a long list of motives and non-negotiable demands that guerrillas might have in stealing nuclear material. Some of these might be (1) to effect long-term changes in political policy, (2) obtain the release of political prisoners, (3) demand the surrender of high government officials as political prisoners, (4) force political resignations, and back a truck up to the treasury. Less plausible but not incredible motives might include obtaining access to classified information or control of intelligence data networks, obtaining large conventional arsenals, occupying a military base, or securing political control over a geographic area.

To achieve its aims, a guerrilla group might use a nuclear hoax. To create an effective hoax, a group might identify a nuclear facility with a large amount of material unaccounted for and then might publicly claim to be in possession of a nuclear device made from the missing material. To make the demand convincing, the group could produce a technically sound bomb design. The hoax would seem even more authentic if accompanied by an actual sample of the nuclear material from which the guerrillas claimed to have made their bomb. Authorities would then have a hard time discrediting the threat.

A much less sophisticated nuclear threat made on October 27, 1970 in Orlando, Florida, almost succeeded. City officials received a letter warning that the entire city would be destroyed by an H-bomb unless a million-dollar ransom were paid, and a safe conduct to Cuba for the terrorist were guaranteed. The letter was accompanied by a convincing sketch of the hydrogen bomb the blackmailer would use. Neither the AEC nor the FBI was able to discredit the threat, and city officials were about to pay the ransom until it was discovered that the instigator of the plot was a fourteen-year-old high school boy without a nuclear weapon.

Guerrillas with a real bomb could send photographs of the device as it was being constructed as well as samples of nuclear material to authenticate the threat. If the demands were not met, the bomb could be delivered to its target in the trunk of a car, by truck, boat, or air. Manufacture of the weapon might be difficult to detect, particularly if a sympathetic foreign country were used as a safe refuge.

Guerrillas also might threaten to contaminate residential or business establishments with plutonium. As noted earlier, if dispersed as an aerosol, plutonium is phenominally dangerous: Placed in the air conditioning system of a large office building, shopping center, or industrial plant, it could expose thousands of people to fatal irradiation. Particles totaling ten thousandths of a gram can cause fibrosis* of the lungs, bringing death within days or weeks.

Even if adequate safeguards were available to make nuclear theft impossible, guerrillas wishing to employ terror tactics would not need to steal special nuclear materials. By sabotaging a nuclear power plant or a reprocessing plant, almost as much, or possibly more, damage could be done than with a clandestine bomb. The risk

---

*The development of excessive connective tissue.

of sabotage was purposely excluded from the Reactor Safety Study of Professor Norman C. Rasmussen because of the alleged impossibility of predicting its chances of occurrence. Yet the means by which nuclear sabotage might occur are virtually unlimited.

According to Dr. L. Douglas DeNike, a technical consultant on nuclear sabotage, by placing conventional explosives at only two locations in a small nuclear power plant, the plant's input transformers could be disabled.[20] This plus sabotaging the emergency diesel generators could cut off all power to the plant's electric pumps. This would disable control rod motors vital for shutdown. In pressurized water reactors, control rods will drop by gravity into the reactor from above, but in boiling water reactors, whose rods are inserted from below, reactor shutdown mechanisms would be crippled by loss of control rod motor operation.

Control of the reactor could also be lost if electrical cables were cut or burned. Large numbers of these cables, with flammable insulation, are bundled together to minimize the number of openings necessary in the reactor containment building. The reactor without its normal pumps is dependent on its emergency core cooling system (ECCS). The ECCS water at some plants, however, is in a virtually unprotected tank sitting outside the reactor containment building. Ordinary explosives could easily puncture it, or damage connections to reservoirs of cooling water.

To illustrate one possible sabotage scenario, the actions of the Humocoms, an imaginary guerrilla group, are described in the next few paragraphs.

THE HUMOCOMS ATTACK

The Humocoms, "commandos for a human world," are a group of highly educated radical technologists. They hold respected positions in society, are well educated, and work clandestinely. The group includes a nuclear physicist, a nuclear engineer, an ex-Navy underwater demolitions expert, and various other ex-military and scientific personnel.

In support of a nationalist movement (perhaps in the Middle East or Northern Ireland), the Humocoms decide on using nuclear blackmail to extort a billion-dollar ransom from the U.S. Treasury, and to demand the withdrawal of U.S. economic and political support from the rulers of a foreign country which the guerrillas oppose.

After meticulous planning, the Humocoms steal a helicopter from the sheriff's department closest to the Multimegawatt I Nuclear Power Station in New Jersey. A few members of the group then approach the plant and conduct a diversionary maneuver near its gates which distracts the plant's two guards. The guards are then pinned down by rifle fire. Meanwhile, the Humocoms descend by helicopter to the plant's roof and force their way inside the locked administration building using small explosives. They carry automatic weapons. Once inside, they force the few employees on duty to surrender. Now in command of the plant, the Humocoms incapacitate the guards and place larger explosives at strategic locations so that the reactor cooling and shut-down systems can be destroyed. The explosives are set with timed fuses and can also be activatated by remote radio signals. With the flick of a switch, the Humocoms can now create a nuclear catastrophe at will by triggering events leading to an inevitable core meltdown.

Because the reactor is located at a remote site connected to main highways by a narrow two-lane road which members of the group have booby-trapped and are covering with rifles, squad cars from nearby police departments are unable to relieve the plant's beleaguered guards, and the National Guard and the Air Force are summoned. But before the military and police can retake the plant, the attackers escape in their helicopter, carrying a portable radio device capable of setting off their explosives. While they escape, other members of their group issue the Humocom's non-negotiable demands:

• No one is to enter Multimegawatt I or the plant will be destroyed automatically through sophisticated electronic detection apparatus connected to the explosives.

• The Humocoms must be assured safe passage out of the country, receipt of a billion dollar ransom, and the surrender of high corporate and governmental officials as hostages, until changes specified by the Humocoms are made in U.S. foreign policy. To show their earnestness, the Humocoms detonate one or two explosives at the reactor, breaching its containment building but not ruining it.

Authorities at this point conclude payment of the ransom is a cheaper option than dealing with a meltdown, so they authorize payment.

*   *   *

Of course, far less complex and more plausible tactics could be used to conduct nuclear sabotage. Moreover, recent events indicate that vulnerable nuclear facilities are already attractive targets for guerrillas. In March 1973, guerrillas seized a nuclear power plant being built in Argentina and escaped after festooning the plant with political slogans and seizing the guards' weapons.[21] A more serious incident occurred one Friday evening in November 1972, when three jittery armed men commandeered a Southern Airways DC-8 jet enroute to Florida and demanded a $10 million ransom. Equipped with guns and grenades and ordering themselves parachutes, the hijackers diverted the plane and the thirty other persons aboard to Detroit, then Cleveland, and ultimately Toronto.

There, they rejected an offer of five hundred thousand dollars and took off again for Knoxville, Tennessee. Before arriving, they threatened to crash into Oak Ridge, the site of an important nuclear facility. Their frightening threat was taken so seriously by officials that all Oak Ridge reactors were shut down and the plant employees were evacuated. The macabre ultimatum was never carried out, but it serves as a stark reminder of the myriad unpredictable human factors to which nuclear power plants and the public are exposed.

Other incidents of sabotage — some by utility employees — also have begun occurring: An arsonist struck the Indian Point 2 plant at Buchanan, New York, in November 1971, causing between five and ten million dollars damage; valves and switches were apparently tampered with at Commonwealth Edison's Zion Nuclear Station near Chicago; severed cables and clogged helium filters were found at the Fort St. Vrain nuclear plant in Colorado.[22] In late summer, 1971, an intruder entered the Vermont Yankee plant grounds, despite its fences and guard towers, and wounded a night watchman before escaping.

And in 1976, it was belatedly revealed by the Nuclear Regulatory Commission that more than one hundred bomb threats have been directed at nuclear power plants in the past several years and that, in two cases, explosives had actually been placed at reactors. Dynamite was discovered outside the University of Illinois, Urbana, research reactor in 1969, and in late 1970, a pipe bomb was found and defused at the 497 MW Point Beach 1 reactor in Two Creeks, Wisconsin, shortly before the plant opened.

*   *   *

As nuclear theft and sabotage become more common, the public should be made aware of the fact that there is virtually no way to eliminate completely the threat of sabotage and stop the dedicated saboteur. No matter how well protected power plants might one day be against ground attacks, for example, they would have to be protected with antiaircraft defenses to prevent bombing from the air. It would also be almost impossible to prevent suicidal fanatics from sending a hijacked plane on a collision course with a nuclear reactor containment building or spent fuel pool. Secretary of Defense James Schlesinger has admitted that, "The nuclear plants that we are building today are designed carefully to take the impact of, I believe, a 200,000-pound aircraft arriving at something on the order of one hundred fifty miles per hour. They will not take the impact of a larger aircraft."[23] A Boeing-747 aircraft weights about 365,000 pounds and can cruise at six hundred miles per hour.

A more vulnerable target is the spent fuel storage area with its enormous inventory of radioactivity. These storage pools are often shielded only by buildings with corrugated steel walls half an inch thick or made of ordinary concrete block. A sabotage team might not want to bother attacking a reactor and might choose the less complex task of dropping a bomb into the spent fuel storage pool. Due to the current absence of reprocessing facilities, many of the irradiated fuel pools are nearly full.

Even many people in the nuclear establishment seem to agree that nuclear sabotage is feasible and would have horrifying consequences. L. Manning Muntzing, former AEC Director of Regulation, conceded that a band of highly trained, sophisticated terrorists could conceivably take over a nuclear power plant near a major city and destroy it in such a way as to kill thousands — perhaps even millions — of people."[24] Former U.S. Navy underwater demolitions officer Bruce L. Welch testified on March 28, 1974 before the Joint Committee on Atomic Energy that, "As one trained in special warfare and demolitions, I feel certain that I could pick three to five ex-Underwater Demolition, Marine Reconnaissance or Green Beret men at random and sabotage virtually any nuclear reactor in the country."[25]

Dr. DeNike also has raised the possibility that these weapons might be used to sabotage the spent fuel casks that carry up to fourteen million curies of radioactivity and travel by rail through the country. "Cleanup of a seriously ruptured cask," wrote DeNike, "might well be impossible to carry out safely save with totally robotized or remote-controlled machinery which does not exist. Continuous release of radioactive [material] from a breached cask on a moving train would contaminate the railbed, very probably necessitating rerouting of the affected right-of-way."[26]

The AEC and the American Nuclear Society have made much of the safety of nuclear fuel shipping casks. These are enormous affairs evidently designed to withstand the rigors of Hell itself. They are meant to maintain their integrity even if dropped thirty feet or exposed to fire for thirty minutes, or submerged for eight hours in water. (The thirty-foot fall is said to be roughly equivalent to the impact on the cask of a typical sixty-mile-per-hour collision.) But would these casks hold up in a thirty-one-foot fall or a thirty-one-minute fire? Or what would happen if a truck carrying such a cask was in a sixty-five-mile-per-hour crash that split the cask and then sent it over a cliff, or off a high bridge into a major river from which it could not immediately be extricated? What if a cask were exposed to the explosion of a railroad tank car carrying hazardous flammable chemicals? Such explosions can be powerful enough to propel large sections of the tank car through the air like a rocket.[27]

Yet of all the containers by which nuclear material is transported, the spent fuel cask is probably the safest from accident and from theft. Thieves could not open spent fuel casks without enormous cranes and underwater holding tanks; to do so otherwise would insure certain death from massive doses of radioactivity. But other more vulnerable nuclear cargo, some containing enough fissionable material for many atom bombs, routinely travel around the country by truck, rail, and air. The crash of an airliner carrying plutonium could cause large-scale contamination; efforts at decontamination would be extremely difficult and expensive and to a large extent fruitless.

Air shipment of plutonium has only recently been banned by Congress after nearly one thousand air shipments of plutonium a year totaling more than twenty thousand pounds have been flown over the U.S. The ban was sponsored by Representative James H. Scheuer (D-N.Y.), who introduced it in June 1975 as an amendment to the Nuclear Regulatory Commission authorization bill.

In recent years, some nuclear materials have been misrouted and others have been lost for days at a time.[28] Because so much is shipped by common carriers, the risk exists that agents of the carrier might conspire to steal the shipment or misaddress it. About 2 percent of everything shipped in this country is stolen[29] and experts on the subject attest that the transportation industry is "so thoroughly infiltrated by the Cosa Nostra that any cargo which organized crime determines to obtain will be obtained. . . . The environment of the transportation industry is one of incompetence, criminality, and unreliability."[30]

Partly because of the unresolved dangers of transporting nuclear materials, nuclear proponents are now suggesting that the building of giant nuclear energy complexes, euphemistically termed "nuclear parks," would help circumvent transportation insecurities. But although locating several phases of the nuclear fuel cycle together in one industrial complex would lessen some risks (such as those of hijacked cargo) by minimizing the need to ship nuclear materials long distances, it would create hideous new ones.

Because these parks may consist not only of power plants but of enrichment, fuel fabrication, and reprocessing facilities, they could create enormous environmental problems and new insecurities. And even nuclear proponents admit that unless these facilities contained the equivalent of ten to thirty large nuclear plants at each site, they would not even be economically competitive with ordinary nuclear electricity.[31] (Based on the data presented in Chapter 7, the writer doubts that *any* proposed nuclear park would be economic.) The size and complexity of these huge facilities would require nuclear parks to be under more or less continuous construction for a decade or much longer. Dr. DeNike has pointed out that due to the sophistication of new bomb detonators, it would be possible for construction or maintenance workers to conceal a bomb "in the concrete or insulation of a reactor's primary coolant loop, to be detonated perhaps years later by automatic timing device or radio signal." [32]

Despite these hazards, a nuclear park project is being actively promoted by the Energy Park Development Group of Harrisburg, Pennsylvania. Backed by four Pennsylvania electric utilities, the group has already identified ten "candidate energy park sites" [33] and has published the layout for "a typical energy park" that would include four reactors and eight cooling towers. Supporters of the concept see it as a way of expediting the licensing of nuclear plants by getting blanket approval in one license to build several plants, rather than being obliged to participate in the arduous process of license hearings and permit applications for each individual plant. Nuclear parks are also being studied by the National Science Foundation, the Nuclear Regulatory Commission, and the Federal Energy Administration. In the AEC publication, *Evaluation of Nuclear Energy Centers*, the implementation of a demonstration nuclear energy center is slated for 1977. [34]

A large nuclear energy center with many reactors would not only multiply the risks of meltdowns and expose the public to larger routine releases of nuclear toxins, the centers would also create new and yet unappreciated difficulties. The concentration of many facilities would produce climatic changes, for example, as many simultaneously operating cooling towers released tremendous amounts of waste nuclear heat and moisture. In addition, the concentration of an entire region's power supplies at one site would expose the whole region to a power failure if the site became unusable because of a natural, accidental, or malevolently caused disaster. A meltdown at one reactor, for example, could contaminate an entire center. Finally a nuclear center would present a tempting military target in wartime.

NUCLEAR PROLIFERATION

As the nuclear industry expands, it increases the risks of nuclear war by actively promoting the sales of commercial nuclear technology abroad to countries that are not now nuclear powers. Reactors sold for civilian power generation or research produce plutonium just like military reactors. Once this plutonium is recovered in a plutonium reprocessing facility, it can be used in an atomic bomb. Many of the U.S. nuclear industry's customers happen to be foreign nations that want nuclear weapons. In some cases, these countries are not even signatories to the Nuclear Non-proliferation Treaty; the treaty is merely a voluntary agreement with no enforcement provision that pledges countries to abstain from acquiring nuclear weapons.

Selling power plants overseas with the ineffectual safeguards against nuclear proliferation that exist today virtually assures that more countries will obtain nuclear weapons. With more nuclear-armed nations, the risk of a war that could annihilate the nuclear industry's profits, and civilization as we know, grows greater every day.

By 1980, about thirty nations will have operable nuclear power plants. Nations with reactors planned or underway include many in the throes of political or military crises, such as Argentina, Brazil, Egypt, India, Iran, Israel, Italy, Libya, South Korea, Taiwan, and Turkey. The fact that so many nations will have nuclear power shows that the floodgates of nuclear proliferation have been opened, and it may already be too late to close them.

Although the U.S. is not the only country responsible for nuclear proliferation, a brief look at the history of this problem shows that the U.S. has literally catered to the needs of insecure, dictatorial, or unstable governments which may believe that nuclear weapons will ultimately increase their security.

The current spread of nuclear technology began with President Eisenhower's "Atoms for Peace" speech to the United Nations General Assembly on December 8, 1953. In that address, Mr. Eisenhower called for the setting up of the International Atomic Energy Agency to redirect nuclear technology from military to peaceful purposes. Until the objectives of the "Atoms for Peace" plan became law, the U.S. was greatly constrained in promoting the use of atomic energy abroad. Under the Atomic Energy Act that created the U.S. Atomic Energy Commission on August 1, 1946, the U.S. was prohibited from exchanging atomic energy information with other nations. But following the "Atoms for Peace" speech, in 1954, Congress passed an amendment to the Atomic Energy Act of 1946. The amendment authorized the United

States to make "the peaceful uses" of atomic energy as available as possible, subject to considerations of defense and security. The next year, twenty-five bilateral agreements were signed by the U.S. with other nations and organizations providing the cooperating countries with nuclear materials for research or power reactors. A subsequent amendment to the Atomic Energy Act in 1964 encouraged the further spread of nuclear power by allowing the AEC to license private industry in the U.S. to own and export nuclear material.

The primary safeguard upon which the world has relied to restrain the proliferation of nuclear weapons as a result of these exports has been the International Atomic Energy Agency (IAEA), which was set up in Vienna in 1957 on the initiative of the U.S. Yet this body has been able to establish only manifestly inadequate controls on the spread of nuclear weapons. It has the power to inspect and inventory nuclear materials, so in theory it is able to detect diversion of nuclear materials to military purposes. But the IAEA has no power to enforce its findings to actually prevent nuclear diversion. Moreover, nations submit to IAEA controls voluntarily under the Nuclear Non-Proliferation Treaty, and the IAEA cannot compel nations to participate in its inspection program. The IAEA itself could be expelled from a country in which it was attempting to perform surveillance.

The agency has an average of less than four inspectors for each of the twenty countries it is legally allowed to monitor, and had a budget of less than $30 million in 1974. Yet it must oversee four hundred thousand pounds of plutonium by 1980 in countries under its formal jurisdiction. Moreover, no international agency at all is regulating the international shipment of plutonium.[35]

Even with such poor international safeguards, the U.S. government has gone so far as to promote the sale of uranium fuel and two 600 Mw nuclear reactors to Egypt and Israel, respectively. Provisional contracts to provide the initial fuel of 115,000 pounds of uranium for each reactor were signed on June 26, 1975, and three U.S. companies — General Electric Company, Westinghouse Corporation and Babcock & Wilcox Company — have already bid on the Israeli reactor contract. Although the U.S. will no doubt insist that neither Egypt nor Israel be allowed to reprocess plutonium from spent reactor fuel, the deal makes plutonium more accessible to these warring countries. Neither Israel nor Egypt is a signatory of the Nuclear Non-Proliferation Treaty. Both countries have delivery capabilities for atomic bombs, and have no special pressing need for nuclear-generated electricity. Egypt today already has more generating capacity from the Aswan dam than it can use.[36] Although both countries have had research reactors since the early 1960s, the reactors offered by the U.S. are many times larger and able to produce vastly more plutonium.*

The U.S. has also helped South Africa toward a nuclear weapons capability. The U.S. recently licensed U.S. Nuclear Corp. of Oak Ridge, Tennessee, to ship the South African regime more than forty pounds of highly enriched (93 percent) uranium-235. The uranium is officially intended to be used in South Africa's research reactor at Pelindaba, the Transvaal. Meanwhile, South Africa is developing its own uranium enrichment plant with the assistance of computer equipment from the Foxboro Corporation of Foxboro, Massachusetts. Because South Africa has an abundant supply of its own natural uranium, it soon will be possible for South Africa to back its policies with threats of nuclear violence.

The nuclear-free zones that once existed in Latin America, the Middle East, Africa, and South Asia are now disappearing. The U.S. has tentatively approved an agreement to ship nuclear fuel to South Korea. The agreement would give Seoul the capacity to make one hundred nuclear bombs if Seoul can retain and refine its reactor plutonium.[37]

Argentina now has a plutonium reprocessing plant as well as an operating nuclear reactor that can produce three hundred thirty pounds of plutonium in its first year. The U.S. has had bilateral agreements for providing nuclear assistance to Argentina and Brazil since 1955. Brazil recently made a four-billion-dollar deal to purchase U.S. nuclear technology from West Germany. The deal provides reactors and equipment for fuel processing, enrichment, and reprocessing.[38]

The U.S. even provided a 250 kW reactor that began operating for the South Vietnamese in Dalat in 1963. Ostensibly the reactor was for experimental research and "to provide power." It is not

---

*As this book went to press, a CIA report was made public indicating that Israel already has 10-20 atomic bombs. New reactors could help add to that arsenal.

clear why the experimental nuclear work had to be carried on in South Vietnam, nor why the experimental results anticipated were considered worth the risk of installing hazardous nuclear technology in an agrarian nation rent by civil war. As the North Vietnamese were about to move into the Dalat area, the U.S. sent a special team to Dalat. After removing the reactor's fuel, the team dynamited the reactor to keep it from being captured.

In all its nuclear Agreements for Cooperation, the U.S. has the right to apply safeguards to insure that its technology is not used for research on, or development of, nuclear weapons. These safeguards can include inspections and bans on plutonium reprocessing by the aid recipient. But no one knows whether these safeguards can or will be adequately enforced. It is especially regrettable that these risks have been incurred by the U.S. to form nuclear partnerships with unstable, repressive, racist, or dictatorial regimes including the Philippines, Spain, Thailand, Taiwan, and Venezuela, in addition to Argentina, Brazil, and South Africa.

While contributing to the growing risk that some foreign government may resort to the use of a nuclear weapon, the U.S. nuclear industry is continuing to promote foreign nuclear sales in troubled areas. Westinghouse and General Electric now supply 70 percent of the world reactor market, and nuclear exports this year are expected to earn more than a billion dollars for U.S. firms.* ERDA has estimated that nuclear plants already on order or recently sold abroad will ultimately bring U.S. suppliers about five billion dollars. Yet although the United States has done so much to create the nuclear proliferation problem and today controls the lion's share of the market, when France announced a bilateral agreement to provide Iraq with nuclear technology, the *Wall Street Journal* reported "the dismay of U.S. authorities who are concerned that the dispersal of nuclear capacity may lead to the proliferation of nuclear weapons."[39] This "dismay" is particularly hypocritical because 45 percent of Framatome (the French seller) is owned by the Westinghouse Corporation, and Framatome sells reactors built with Westinghouse technology. Kraftwerk Union,

---

*The U.S. is not the only nation engaged in hawking nuclear materials. Other suppliers are Britain, Canada, France, Sweden, the USSR, and West German, although Britain, Sweden, and the USSR are not particularly active in nuclear exports.

the West German nuclear exporter, is also peddling U.S. technology through licenses obtained both from Westinghouse and General Electric.

Eighty percent of all U.S. reactor exports are financed by loans from the U.S. Export-Import Bank (Eximbank), a wholly owned government corporation that provides loans, or loan guarantees, to promote the export sales of U.S. goods and services. Through the end of 1974, the Eximbank loaned or guaranteed more than three billion dollars worth of U.S. reactors and fuel to eleven countries. Eximbank loans are usually at 8 percent interest for fifteen years and cover 45 percent of purchase costs while offering guarantees of private financing for an additional 40 percent. According to the *New York Times,* the Eximbank is authorized to commit $3.4 billion a year to encourage nuclear exports, and this program, in the words of Eximbank President William J. Casey, "rates a very high priority."[40] The nations which have been granted Eximbank assistance are: Brazil, France, West Germany, Italy, Japan, South Korea, Mexico, Spain, Taiwan, Sweden, and Yugoslavia.

In a $96,000 study done for the Energy Research and Development Administration in early 1975, the Washington consulting firm of Richard J. Barber Associates concluded that the U.S. government should not be encouraging the sale of nuclear materials to the less industrialized nations. Expressing concern about the thousands of pounds of reactor plutonium to be produced each year, the study found "that existing international controls are inadequate and *unlikely to grow stronger*"[41] [emphasis added]. Because of the size of exported U.S. reactors and their unreliability, as well as for other reasons, the Barber study declared it would not be in the economic self-interest of many less industrialized countries to buy U.S. reactors now.

Apologists for the sales of nuclear technology abroad have argued that if the U.S. does not promote them, other countries will sell the nuclear materials anyway, possibly with worse safeguards than our own. This policy is illogical and unsound. Instead of using others nations' folly to excuse our own, the U.S. should set a better example by halting its nuclear exports and intensifying its efforts for world nuclear disarmament. The nuclear proliferation that has been occurring in the last thirty years is proof that we have not made enough efforts to control the nuclear technology which we pioneered and for whose proliferation we bear so much responsibility.

The Nuclear Regulatory Commission in April 1975 temporarily stopped processing licenses for nuclear reactor exports and for large exports or imports of nuclear materials, pending a review of export policies. But a draft environmental impact statement of the nuclear export program issued by the Energy Research and Development Administration in August contains strong preliminary indications that the government plans to continue promoting nuclear exports.

In addition to stopping nuclear sales abroad and terminating existing contracts, the U.S. should renounce domestic plutonium reprocessing and consign all plutonium-bearing spent fuel rods to storage, rather than continuing to spread this carcinogenic nuclear weapons material all over the world. A multinational nuclear fuel reprocessing facility, such as suggested recently by Secretary of State Henry Kissinger, is not an acceptable solution to the problem of plutonium proliferation. It would, instead, legitimate the proliferation of environmentally unsound nuclear energy economies throughout the world. And it would institutionalize rather than eliminate nuclear environmental problems and terror/sabotage risks.

The outlook for a major reassessment of nuclear foreign policy is not a cheerful one. Several U.S. administrations have used nuclear reactors as a way of buying the favors of foreign nations. The Ford Administration policy is to treat nuclear technology as an article of commerce and sell it, commercially, while strengthening safeguards. This is ideal for the U.S. nuclear industry which anticipates multibillion-dollar foreign nuclear sales. It is not ideal for world peace, but there is no economic profit for the nuclear industry in renouncing the nuclear export business, and there are many political hurdles to its renunciation.

The only sane policy would be to couple a U.S. export moratorium with ironclad international safeguards against proliferation from other sources, while trying, by international agreement, to halt the spread of reactors abroad. It is unrealistic to expect an administration such as the Ford-Rockefeller government to help stem nuclear expansion abroad while the administration has exerted every effort to promote nuclear power domestically. Reform of domestic energy policy, therefore, appears to be a necessary but not sufficient prerequisite to ending nuclear proliferation abroad. And the only way to alter domestic policy dramatically is to expose the U.S. government to a storm of protest over nuclear reactors and the nuclear industry's shenanigans.

NOTES

1. David M. Rosenbaum, *A Special Safeguards Study,* Report to AEC Director of Licensing John F. O'Leary. Other participants in the study were described by Senator Abraham Ribicoff when he released the document and entered it in the *Congressional Record* (Senate, Vol. 120, No. 59, April 30, 1974): "Dr. John Geogin, Union Carbide Nuclear Division; Robert Jefferson, Sandia National Laboratory, AEC; Dr. Daniel Kleitman, professor of mathematics, MIT; William Sullivan, former Assistant Director of the FBI and former Director of the Office of National Narcotics Intelligence; and Dr. David Rosenbaum, consultant on terrorist threats to technology and former assistant to Mr. Sullivan at ONNI."

2. Mason Willrich and Theodore B. Taylor, *Nuclear Theft: Risks and Safeguards,* A report to the Energy Policy Project of the Ford Foundation (New York: Ballinger Books, 1974), p. 108.

3. *Congressional Record*, 93, Senate, Vol. 120, No. 59, April 30, 1974, p. 6621.

4. *A Time to Choose, America's Energy Future,* Energy Policy Project of the Ford Foundation (New York: Ballinger, 1974).

5. Willrich and Taylor, *op. cit.,* pp. 66 and 68.

6. "Theft of Missiles Confirmed," *San Francisco Chronicle* based on *United Press* dispatch from Heidelberg, September 1974.

7. "Terrorists and A-Arms in Europe," *San Francisco Chronicle/United Press*, September 26, 1974.

8. *Science*, April 9, 1971, as cited by John P. Holdren, *Energy* (San Francisco: Sierra Club, 1971).

9. *The Nuclear Power Alternative*, Special Report 1975-A, January, 1975.

10. John McPhee, *The Curve of Binding Energy* (New York: Farrar, Strauss and Giroux, 1974).

11. Willrich, as quoted in McPhee, p. 127.

12. "U.S. Says Lost Plutonium Is 'Only a Small Amount,'" *New York Times*, January 3, 1975.

13. U.S. ERDA, *U.S. Nuclear Power Export Activities,* environmental statement (draft), August 1975, Appendices C-3 and C-4.

14. "A.E.C. Orders Utilities to Arm Guards," *New York Times*, May 1, 1974.

15. "3,600 Lost Nuclear Jobs in Year, Many to Alcohol, Drugs," *Los Angeles Times*, January 27, 1974, as quoted by L. Douglas DeNike, "Unacceptable Security Deficiencies in the Draft EIR, Proposed San Joaquin Nuclear Project."

16. Russell W. Ayers, "Policing Plutonium: The Civil Liberties Fallout," *Harvard Civil Rights-Civil Liberties Law Review,* v. 10, no. 2, Spring 1975.

17. Reported in "The Plutonium Connection," television program in "Nova" science series, produced by WGBH, Boston, aired March 9, 1975 by Public Broadcasting Service.

18. Willrich and Taylor, *op. cit.* pp. 163-1964.

19. David Krieger, "Terrorists and Nuclear Technology," Monograph, Oct. 21, 1974.

20. L. Douglas DeNike, "Malice in Atomland at San Onofre," Monograph, 1975.

21. Krieger, *op. cit.*

22. L. Douglas DeNike, "Unacceptable Security Deficiencies in the Draft EIR, Proposed San Joaquin Nuclear Project," Statement submitted to Los Angeles Department of Water and Power, June 1975.

23. Krieger, *op. cit.*

24. DeNike, "Inadequacies of Nuclear Power Plant Security Systems: Some Shocking Truths," Statement to State of Washington Thermal Power Plant Evaluation Committee Pursuant to Security of the Proposed Skagit Nuclear Power Project, June, 1975.

25. *Ibid.*

26. *Ibid.*

27. Eric Albone and Julian McCaull, "Freighted with Hazard," *Environment*, vol. 12, no. 10 (December 1970).

28. Environmental Action Reprint Service, "And Now for a Little Diversion," *Stockholm Conference ECO*, vol. 2 (September 8, 1972).

29. McKinley C. Olson, "The Hot River Valley," *The Nation,* August 3, 1974.

30. Samuel Edlow, consultant on the transportation of nuclear materials, April, 1969 meeting, Institute of Nuclear Materials Management.

31. Willrich and Taylor, *op. cit.,* p. 69.

32. DeNike, "Unacceptable . . . "*op. cit.*

33. "Energy Park, A Study on Meeting Future Needs in Pennsylvania," Energy Park Development Group, 301 APC Building, 800 North Third Street, Harrisburg, Pennsylvania 17102. The four utilities in the EPDG are Pennsylvania Power & Light Company, Philadelphia Electric Company, Pennsylvania Electric Company, and Metropolitan Electric Company.

34. U.S. AEC, *Evaluation of Nuclear Energy Centers*, Executive Summary, WASH-1288, Vol. 1, January 1974. Now available from U.S. Nuclear Regulatory Commission.

35. "The Holes in Atomic Inspection," *San Francisco Chronicle,* August 21, 1975.

36. *San Francisco Chronicle,* June 18, 1974.

37. David Burnham, "New Curbs Urged on U.S. Nuclear Sales Abroad," *New York Times,* April 24, 1975.

38. "Brazil Atom Deal Worries Senators," *San Francisco Examiner*, June 26, 1975.

39. "France to Give Iraq Nuclear Technology and Atomic Reactor," *Wall Street Journal*, September 15, 1975.

40. Ann Crittenden, "Surge in Nuclear Exports Spurs Drive for Controls," *New York Times,* August 17, 1975.

41. David Burnham, "Study Asks for Caution in Sale of Atom Reactors," *New York Times,* August 20, 1975.

Part II

CLEAN ENERGY ALTERNATIVES

10

THE ENERGY CONSERVATION TREASURE

By 2010, conservation could save an amount [of energy] equal to our present usage, making it the most important step we can take. For the most part, energy-efficiency investments in all sectors are the least expensive new sources of energy we will ever find.

— Roger W. Sant, Assistant Administrator,
Federal Energy Administration in a letter
to the *New York Times,* October 17, 1975

The U.S. has a massive hidden energy reserve equal in value to billions of barrels of oil, trillions of cubic feet of natural gas, or hundreds of billions of tons of coal. This energy has lain untapped for years. Yet to benefit from it, we don't need to invent any fundamentally new energy technology, we don't need to construct vast numbers of power plants, we don't need to ravage the environment, and we don't need to extract energy from expensive marginal sources such as oil shale and tar sands.

The energy bonanza is conservation. Its potential is so enormous that — if tapped extensively — it makes the building of uranium fission plants completely unnecessary.

HOW MUCH ENERGY DO WE NEED?

For purposes of energy analysis, a convenient way to grasp the magnitude of the U.S. energy conservation potential is to convert all energy used in society — whether the energy is made from oil, coal, gas, or hydroelectric sources — into a common unit, the British thermal unit (Btu). Every finite energy source can be expressed in Btu's; one Btu equals the amount of energy it takes to heat a pound of water one degree Fahrenheit. A forty-two gallon barrel of oil, for example, equals 5.8 million Btu. A ton of Eastern coal equals 26 million Btu. A thousand cubic feet of gas equals a million Btu. For convenience, very large numbers of Btu are expressed in quad units (q), each equal to a million billion Btu (10^{15} Btu, in scientific notation). When the Ford Foundation Energy Policy Project added up all the energy used in the U.S. in 1973, it arrived at a figure of about 75 q.*

However, not all this energy is really necessary for our social or economic welfare. Dr. George Kistiakowsky, a former science adviser to President John F. Kennedy, attested recently that "a 30 to 40 percent reduction in energy use is entirely feasible."[1] Many other knowledgeable scientists support this view. Physics professor Marc H. Ross, at the University of Michigan, and Dr. Robert H. Williams, senior scientist for the Ford energy project, indicate that, over time, 34 q or *45 percent* of U.S. energy use could be saved by a comprehensive energy conservation effort, not counting the energy cost of the conservation measures themselves.[2]

Imagine that with the wave of a wand, all the conservation

---

*This 75 q , although purportedly a measure of total energy use, actually is only a partial tally concealing a strong hidden bias against solar power. The energy supplies counted in the 75 q are fossil fuel power, nuclear power, hydro power, and geothermal power. The Ford Foundation Energy Policy Project lists the nuclear contribution in 1973 as 1 q. The project does not list *any* contribution due to direct solar energy. But as Steve Baer pointed out in *The Elements* (November 1975), the nuclear contribution is minuscule compared to the direct solar energy used, for example, in growing food for the nation. Solar energy also makes another tremendous but uncounted contribution for warming houses that stand in the sun, for turning windmills, and for drying clothes on lines. If an energy pie (circle graph) were drawn showing the relative importance of direct solar energy compared to all other sources combined, solar would be the largest slice in the pie by far.

measures suggested by Ross and Williams had already been implemented as of 1973. Energy use would have fallen so low in that year that, even with the rate of energy growth at the (high) pre-oil-embargo level, the U.S. would not have reached *real* 1973 levels of energy consumption until the mid-1990s! All that time would have been available to develop improved solar power technologies and other clean energies before energy use reached current rates.

The 34 q energy savings projected by Ross and Williams may well be an underestimate. The two scientists did not consider any conservation measures that would require lifestyle changes, nor did they project the energy savings that may be possible with new energy technologies that are likely to become available between now and the year 2000. Instead, Ross and Williams confined their estimates to forecasts of the potential impacts of current or imminently available commercial technology, although technology now evolving will revolutionize energy use, making phenomenal fuel savings possible.

On the horizon are decentralized energy-use systems based on the integrated use of thermal batteries, heat pumps, and *external* combustion engines, such as the Stirling engine. These systems could be linked to solar, wind, geothermal, or conventional generating devices with resulting overall fuel efficiencies that are fantastic by conventional standards. Such self-contained energy systems will ultimately be capable of cutting system fuel requirements to zero. Use of these systems will make a mockery of all but the most visionary energy-need projections available today, and pilot production units for the home are only two to four years away. But because these comprehensive energy systems are still not commercially available, and because their availability is not essential to the case against fission power, we will first discuss the energy potential realizable from conventional forms of energy conservation using available energy systems. The more advanced thermal battery/Stirling engine option is discussed later in this chapter under the sub-heading, "Decentralized Energy."

If we express all energy in thermal Btu's, we can compare the energy to be saved by conservation with the energy to be had by expanding energy supply options, such as nuclear power. If the 34 q of energy that could be saved by conservation were all converted to electricity, at normal power plant efficiencies, that would save as much electricity as 680 nuclear power plants could produce. This would eliminate the "need" for those 680 plants, assuming one believed such a need existed. Each plant operating at an average efficiency of about 30 percent can produce 1.6×10^{13} Btu(e). So 680 plants can produce about 11 q of electricity, which is about a third of 34 q. Of course, not all energy which is conserved represents a *net* energy savings. Just as some of the energy produced by a nuclear plant must be subtracted from its gross output to account for the energy consumed in building the plant, fueling it, caring for its wastes, and eventually decommissioning it, so some energy is used in the conservation process itself. For example, in building more efficient machines that use less fuel or electricity than those machines now use, a "first cost" must be paid for the machine in energy as well as in dollars. But once the efficient machine begins operating, the resulting energy saving begins paying back the initial "energy debt" incurred in its design and manufacture. This repayment continues year after year, as long as the machine continues to operate efficiently. Even when the energy necessary to implement energy conservation in the U.S. is subtracted from the total energy saved, the energy bonus is still impressive.

The only alternative to conservation is new energy supplies. But to expand supplies while neglecting the ever-growing stock of inefficient machines and houses, where that energy is squandered, is exactly like pouring more and more water into a leaky bucket. The cost of most one-time energy conservation measures usually is lower than the aggregate cost of surplus energy used year after year throughout the lifetime of inefficient equipment. Naturally, a cost-benefit analysis needs to be made in each specific instance, but for the conservation steps recommended in this chapter, it is generally cheaper to save a kilowatt of electricity than to raid nature's storehouse for uranium, or coal, and then consume it in a power plant with a thermal efficiency of less than 30 percent for uranium.* But this is precisely the kind of simple insight that the nuclear power vendors don't want people to gain.

Consider first the size of the U.S. energy budget. On the

---

*Thermal efficiency measures the amount of a fuel's energy that is actually recovered as useful work or heat. For most electricity sources, roughly 70 percent of the heat generated is unavailable to consumers as electricity by the time they put a plug into an outlet.

average, each American consumes six times as much energy per person as people in other countries.[3] Although we are only 6 percent of the world's population, we use nearly a third of its energy. Compared just to other industrialized countries, our per person oil consumption is two to three times as high as theirs. Per person, this energy gluttony absorbs considerably more energy annually than can be had from three thousand gallons of gasoline. Yet for years representatives of the energy complex have clamored in their ads for us to use more energy.

Now they contend, unconvincingly, that major reliance on energy conservation to meet energy demand is tantamount to economic stagnation. Switzerland and Sweden use less energy per person and have higher standards of living than the U.S., as measured by their gross national products (GNP).*[4] Sweden, incidentally, recently postponed eleven of thirteen planned nuclear plants.[5] Denmark, using roughly half our per capita energy, has a per capita GNP equal to the U.S. Our energy use, too, could be substantially cut without adverse economic effects. The prosperous Scandinavians are not having economic paroxysms because of their careful energy use. They're just using energy more efficiently than we are, and their lifestyles are somewhat more energy conserving, as well. Conservation based on efficiency can *increase* rather than decrease a country's GNP.

Energy use and economic well-being are not linked in a one-to-one relationship. As Earl Cook pointed out in *Scientific American*, U.S. energy use rose 50 percent from 1900 to 1920,† yet GNP did not rise at a comparable rate.[6] Conversely, GNP rose from 1800 to 1880 in the U.S. while per capita energy use fell.

Because of the tremendous energy savings possible through conservation and the implicit opportunity for consumers to avoid purchasing additional supplies of energy, the energy complex has used its public relations machines to confuse the public about what conservation really means.

---

*GNP is by no means an accurate yardstick of social welfare: It can include production geared for war, and it fails to reflect factors like economic inequality, injustice, pollution, and stress. But because GNP is a readily available and widely accepted standard of national welfare, we'll continue using it here for lack of anything better.

†The rise occured because of the introduction of the passenger car and the substitution of electricity for more efficient direct uses of fossil fuels.

Much electric utility advertising over the past few years has created the impression that a failure to greatly expand energy supplies means Americans may have to do without light, power, dishwashers, or washing machines. This tacitly dismisses energy conservation's immense potential contribution to our stock of available energy. Moreover, because Americans have become so dependent on electricity and labor-saving appliances, threats to curtail electricity create irrational emotional support from distressed householders for utility sales efforts. This inveigles the already victimized consumer into endorsing unnecessary and expensive utility expansion that leads to his or her further exploitation.

OPTIMIZING ENERGY USE

In reality, energy conservation does not cause valid energy needs to be ignored or denied. It simply means increased energy efficiency instead of shortages and it means *optimized* energy use, so we can enjoy greater well being from whatever energy is supplied. Three types of conservation will now be considered.

One way to conserve is to reduce energy demand by improving the design of an energy use system. For example, adding insulation to a home so less heat energy is required to raise rooms to desired temperatures alters the end use system by a "leak plugging" strategy.

A distinctly different conservation approach reduces energy need by improving the efficiency of mechanical devices used for converting energy to a more useful form. The Franklin stove, in which the energy in wood is changed to heat, needs less wood than a fireplace to make a house just as warm. An air conditioner with a higher performance rating provides more cooling for less power. Both devices enable us to do more with less, not to do without.

Neither of the two basic conservation approaches above necessitates changes in lifestyle, although potentially large energy savings often can be had in exchange for relatively slight changes in the way we do things. In geographic areas where recycling is practiced, bundling up used newspapers may mean spending a moment tying the papers and leaving them at a collection point. The inconvenience is usually trivial, but on a societal basis can result in huge timber savings.

Recycling of metals has an especially large energy payoff. Economics professor G. A. Lincoln estimates that for non-ferrous metals, recycling requires only a fifth as much energy as does primary processing.[7] And the production of primary metals is one of the six manufacturing activities that use half of all energy consumed in U.S. industry.*[8]

Opting to use large appliances or power tools early in the morning or on weekends when power demands are lower entails a lifestyle change that is helpful in a different way. By tending to lower a utility company's peak power demands, the consumer enables the utility to avoid installing extra power capacity. Significant percentages of utility capacity are idle except when needed to meet peak demands. By leveling the peak — spreading the power use to times when demand is less — fewer plants are needed in the system.

A more difficult change for consumers to make is to renounce our cars for short neighborhood trips, using feet, bicycles, or public transport, which gets more passenger miles per gallon of fuel consumed. Both the leveling of peak loads and the shift to public transport from private cars are "belt-tightening" conservation approaches.

Sometimes the line between energy conservation and the use of alternative energies becomes blurred. Alternative energies such as the utilization of urban refuse for heat and power (see Chapter 12) and the use of agricultural waste for methane generation are really forms of energy conservation. Their contribution to our power supply can be of great importance. Farno Green in the G.M. Technical Paper, *Agricultural Wastes,* has suggested that utilizing two-thirds of U.S. agricultural residues would provide as much energy as a third of U.S. coal production! Unlike the coal saved, however, the agricultural wastes are renewable, as long as Americans eat food and raise livestock. (Other alternative energy sources are discussed in Chapters 11 and 12.)

Obviously, none of the three kinds of conservation strategies mentioned entails direct economic cutbacks to any but the energy industry itself, and cutting back there could have a stimulating effect on the rest of the economy. A hundred dollars *not* spent in

---

*The others are chemicals; coal and oil; stone, clay, glass; paper; and food.

the energy sector could be used in other economic sectors. And studies have shown that for each dollar invested in the highly mechanized energy sector, only a relatively few jobs and only a relatively small increment in GNP result.[9] Therefore, a dollar invested in manufacturing, commerce, or service industries instead of in the energy sector, on the average, *must* produce more jobs.

THE ENEMIES OF CONSERVATION

One excuse used by the energy promoters to belittle energy conservation as a major option is the "burdensome" energy demands of pollution control and cleaning up the environment. However, the National Petroleum Council, has estimated that only 4 percent of U.S. energy in 1980 will be needed for pollution abatement in that year.[10] The case against large-scale energy conservation rests on bluff, often from the people with vested interests in nuclear power. The energy industry has based its claim that nuclear power is necessary on simplistic projection of "energy demand." Many of these projections assume —

(1) that energy use will grow as fast in the coming quarter century as during the eight years from 1965 to 1973, when energy use was expanding at 4.5 percent a year and electricity use was growing at 7 percent;[11]

(2) that large-scale energy conservation will not be employed.

These assumptions seem increasingly dubious. For example, from April 1974 to April 1975, the Federal Power Commission found that electricity consumption grew not by 7 percent, but by half a percent. And during the first twenty weeks of 1975, electricity output grew at only 2 percent — much less than the "historical rate" — apparently due to increases in utility rates. The energy combine "forgot" one important fact: as energy prices rise, people cut back on consumption. Based on the 7 percent rate of increase in electricity, use, the energy industry published estimates showing electricity consumption doubling every ten years. That would require the creation of a monstrous electric generating capacity, more than 400 percent larger than the capacity of all the U.S. plants in operation in 1975. So many new plants would be needed that just siting them would be a tremendous problem. If the plants were built, their pollution and waste heat would be gigantic. But this folly can be avoided. The Ford Foundation Energy Policy Project report found that energy growth in the U.S. could be reduced to zero before 1990, and that zero energy growth could be sustained thereafter. Known as the Zero Growth Scenario, this option would require no major expansion of nuclear power. [12] Moreover, by the year 2000, this growth strategy would allow GNP to double while creating more jobs than the higher energy consumption policies studied by Ford.

But the energy complex continues to assert that conservation means economic contraction — a world of austerity and fuel rationing. In its house organ, *Orange Disk,* Gulf Oil Company stated under the heading, "Management's View," "The truth is that conservation cannot do the job by itself. An important fact is the economic consequences of conservation. Many seem determined to achieve energy reduction by means of government-enforced rationing or allocation, or by imposing oil import quotas."*[13] Fortunately, extremely effective conservation can be implemented without any government-imposed rationing whatsoever. The choice between Gulf's policies and rationing is a false dichotomy.†

But Gulf is not likely to make windfall profits from energy conservation, so the thrust of their policy is to encourage the expansion of supplies. "The main reason why conservation cannot stand alone," the Gulf statement added, " . . . is that, for all its helpfulness, conservation does not produce energy. And the United States must not only save energy but achieve a quantum jump in domestic energy output if the nation is to ensure its political security and economic survival."

Here the public is threatened with political disaster and economic difficulties unless reliance on coal and fission is greatly expanded. But domestic oil production simply cannot take "a quantum jump" upward. Most of the economically recoverable oil and gas in the U.S. has already been found and, despite intensifying exploratory efforts, the returns per dollar invested and per

---

*Giant oil companies like Gulf gained large profits from the inflated oil prices made possible by oil import quotas. Yet here, Gulf cavalierly refers to quotas as though they were an invention of conservationists, not something for which Big Oil fought, in order to jack up domestic oil prices.

†The second half of this chapter describes how energy consumption can be reduced without rationing or other extreme measures.

foot of drilling are falling. Energy production from domestic oil and gas peaked several years ago, and annual production has been declining ever since.

The dire warnings of the oil industry have been echoed by the natural gas industry, recently accused of withholding production while crying "shortage," in order to boost gas prices.[14] In hearings on energy policy in 1974 before former Interior Secretary Rogers C. B. Morton, Henry Linden, president of the Institute of Gas Technology, urged that the public be warned about the dangers of conservation. Tell them, said Linden, that cutting U.S. oil consumption even 1 million barrels a day* "means austerity; it means a drop in real income, it may mean unemployment; it means a loss in mobility, and many other undesirable side effects."[15]

Such energy blackmail makes energy conservation seem a particularly ominous threat, yet as noted earlier, spending for energy efficiency is consistent with fuller employment and a higher rate of growth. The problem for the energy combine is that conservation reduces its economic power and influence. No matter how many dollars are spent on energy conservation by society, the spending would be diffused throughout the economy for more efficient machines and structures. It would not go straight into the energy industry's coffers.

WHAT CAN BE DONE

Major energy conservation measures can be implemented in many areas of the U.S. economy, biting deeply into the energy shopping lists that consumers would otherwise have to take to the energy complex. Savings can occur in each of the four major energy-consuming sectors of society: transportation, industry, commerce, and the home.

The largest user of energy in the U.S. economy is industry, consuming more than 40 percent of the total. Waste abounds in this sector. Responding to the oil embargo and higher oil prices, many industries voluntarily instituted energy conservation programs in 1974. The results were encouraging: IBM Corporation was able to save 30 percent of its fuel in thirty-three major plant, laboratory, and headquarter facilities.[16] J.C. Penney Co. and Sears, Roebuck

---

*Out of a current 17 million.

& Co.'s distribution center in Atlanta did as well. Other firms reported far higher savings. Yet by no means did these measures involve more than "belt tightening" measures. The companies just eliminated waste by turning off lights and equipment when not needed, and they reduced unnecessary illumination. Thermostat settings were lowered to sixty-eight degrees. No major investments by the companies were required. No impairments in plant operations resulted. And no significant lifestyle changes were required of employees, except perhaps to put on a sweater or turn off a light occasionally — no austerity was needed.

Undoubtedly many other enterprises could achieve significant energy savings by more careful plant management, especially by the correct adjustment of combustion equipment and by the use of heat recovery systems. In many industrial furnaces, half or more of the heat generated escapes through the chimney, yet devices often could be used to recapture some of this heat and put it to use in the plant.

Improvements in the efficiency of industrial fuel use do not necessarily require long time lags for implementation. Engineer Charles A. Berg, a Federal Power Commission expert on industrial conservation, has pointed out, "Far from requiring fifteen years or more for significant changes to be brought about, industrial production can be revolutionized in less than a decade if the incentives are sufficiently strong." [17]

Currently, U.S. industry consumes about 17 percent of the nation's fossil fuels to make steam. A recent National Science Foundation Study coordinated by the Dow Chemical Company found that if industry would use some of that steam to make electricity, this cogeneration of steam and electricity would save the equivalent of 680,000 barrels of oil a day — by 1985 — 4 percent of our current daily oil consumption. [18,19] That is roughly equivalent to the electrical output of thirty large (1000 MW) nuclear power plants, more than half the total number in operation today in the U.S. That certainly is a lot of power to be saved from just one conservation measure.

Cogeneration is not a new or an untried process. In the 1920s and 1930s, major paper companies found it very profitable to generate electricity with steam produced for paper pulping and other industrial processes. The firms were able to produce three to four times as much electricity as they could consume, but had to abandon its production after the U.S. Justice Department took steps that required the companies to choose between paper and power production.[20] Cogeneration once again could become an important source of power.

In the transportation sector of the economy, which uses about a quarter of all our energy, enormous savings are possible. The average American car on the road now gets about fourteen miles per gallon and its engines use up about 13 percent of all the energy consumed in the nation, yet cars are already on the market that can deliver two and in one case three times that mileage. Dr. Edward Teller, a prominent physicist and strong nuclear proponent, has claimed, " . . . except for the mental inertia in Detroit, we could have a fifty-mile-per-gallon car in perhaps as short a time as three years." Whether the cause of Detroit's sluggishness is mental inertia" or reluctance to retool because of profit considerations, it is clear that doubling the average mileage of all cars on the road would result in a fantastic energy savings of 6 percent (ninety nuclear power plants), without even requiring alterations in travel modes. Additional savings can be achieved by encouraging more people to shift from reliance on one-occupant passenger cars to carpools, buses, and trains.* All these modes of public transport require far less gas or oil per passenger, and they reduce ambient air pollution. Considering that the fleet of cars now on the road will be junked anyway by 1985, it would be far simpler to induce people to make their next car an energy-efficient one rather than to build more fission plants, if the country is in such desperate need of more energy.

Car buyers would gain in this transaction because their cars would be cheaper to operate, and consumers of electricity would gain because they wouldn't be burdened by paying off expensive capital investments in nuclear fission plants. If foot-dragging from the auto lobby and the energy lobby could be surmounted, the shift to smaller cars could be facilitated by putting a federal tax on energy inefficient cars. This revenue could be used to partially

---

*Railroads are four times as efficient as trucks for intercity freight transport and sixty times as efficient as planes in terms of the number of Btu's they require for every ton-mile.[21]

defray the costs of possibly providing a subsidy to small-car purchasers.[22]

Another way to reduce gas consumption would be to add methanol to stretch gas supplies. Methanol can be used to replace 15 to 30 percent of a car's gas supply, with apparent increases in fuel economy and decreases in carbon monoxide pollution. Some methanol is likely to be made from urban and agricultural refuse in the future, but additional supplies would probably have to be made from coal, so methanol is not an ultimate solution to reducing energy demand.

Much energy can also be conserved in U.S. homes, which now account for 19 percent of energy consumption. Pennsylvania Power & Light Co. has been experimenting with a $120,000 model experimental energy conservation house near Schencksville, Pa. The project's director expects the effort will result in devices enabling average homeowners to save two-thirds of their home energy use. The energy savings would come from a number of systems that are now being tested in the home, including a solar-powered water heater, a heat pump, and heat exchangers.* The solar water heater simply uses the sun's energy to heat water, which is stored for later use. Throughout the Pennsylvania Power and Light house, heat exchangers are used to recover waste heat from appliances and even from used drain water.

No matter how promising the preliminary results from the energy conservation house look, the projection that two thirds of residential energy use might be saved requires qualification. This achievement will probably take years, and even when these home conservation devices are economic to install in new homes, the extra cost of adding new heating and cooling devices to *existing* housing, which already has working systems, may be prohibitively expensive. Nonetheless, it is clear that tremendous energy savings

---

*A heat pump operates on the same principle as a refrigerator or air conditioner. A volatile fluid is successively evaporated and condensed again into a liquid by a motor or gas-operated absorption unit. When the liquid is evaporated, it gets cooler and will absorb heat from its surroundings. Compressing the fluid heats it, and it can then transfer its heat to a new surrounding. It can be used for heating or cooling. The heat pump's great advantage is that electric models are 2.5 to 6 times as efficient as electric resistance heaters.[23] Power is not used to generate heat, just to move it from one place to another.

are possible in the residential sector, where more than half the energy used (57 percent) goes for often-inefficient space heating.[24]

Currently, most furnaces are only 75 percent efficient as sold, and haphazard maintenance often cuts their performance to between 35 percent and 50 percent efficiency.[25] Electric resistance heating is even worse. Although electricity available at the wall socket is almost 100 percent efficient in its conversion to heat, that ignores the important fact that for every unit of electric power produced and available at the socket, three units of fuel must be burned in a power plant and another 10 percent or so of the resulting electricity is lost in power transmission. Consequently, the efficiency of electric resistance heating is less than 30 percent. Nonetheless, utilities have promoted the electrically heated home as a symbol of glamorous modern living, whereas it really entails the old-fashioned waste of our fuels, saddling homeowners with staggering electricity bills.

Not only are the heaters in homes and businesses often inefficient, but the buildings themselves leak enormous amounts of heat because of inadequate insulation and excessive drafts around doors, windows, and chimneys. U.S. structures typically have four times as much air infiltration as needed, and this accounts for 25 to 50 percent of the building's heating and cooling requirements. Better construction results not only in smaller heating bills, but in lower air-conditioning costs.[26] Writing in *Science,* Charles A. Berg has estimated that by installing adequate insulation and storm windows in existing homes in accordance with the 1972 Federal Housing Administration minimum property standards, homes built before the standards were in force could cut their fuel needs by 40 percent.[27] Berg believes that similar savings could be had in the commercial sector as stores and offices do not differ greatly from homes in their insulating and ventilating characteristics.

A major study of energy efficiency in buildings by the American Institute of Architects has concluded that by thorough insulation and other efficiency measures in both old and new structures in the U.S., more energy could be saved by 1990 than could be produced by either domestic oil resources (including oil shale), or by a great expansion of nuclear power. More than twelve million barrels of oil a day could be saved, said the AIA, as against eleven million barrels of domestically produced oil.[28]

Simple calculations show that it would take about five hundred large nuclear power plants to save those twelve million barrels of oil, if there were enough high-grade uranium to fuel the nuclear plants. And construction of those plants would require at least $500 billion, plus additional tens of billions for fuel enrichment and waste management plants.* Reprocessing, if federally approved, would require the construction of still more facilities. The AIA found that although the new building standards would be expensive, they would be cheaper over the buildings' lifetimes, than expanding fuel supplies, although the AIA found the initial cost would be higher. The environmental costs of the conservation approach, however, would be decisively less than expanding the energy supply.

EXTRA DIVIDENDS FROM CONSERVATION

Since conservation would mean that less fuel would have to be supplied, less prospecting, less oil drilling, and less strip-mining would be necessary. This would mean less offshore oil drilling, fewer oil spills, fewer supertanker terminals, fewer pipelines, fewer refineries, fewer dusty carloads of coal on the rails, and fewer coal- and oil-storage areas. Dispensing with these developments would create "second order" environmental benefits. For example, reducing the amount of mining would save scarce water needed for western agriculture, and would avoid disruptive water-diversion projects. Less mining and drilling would slow the industrialization and commercialization of rural resource-extraction areas, bringing them less population from urban areas and fewer "blessings" of urbanization such as increased crime, social ills, and loss of rural character.

With less energy extraction going on and less induced industrial development, fewer roads, factories, and trucks would be needed. Naturally, less fuel would be combusted on the road and in power plants, so there would be less air pollution and less ash for disposal from plant sites. And less land probably would be needed

---

*The cost of enrichment plants alone from 1975 - 1990 may exceed $30 billion, according to *Business Week* (November 17, 1975). And according to a *Wall Street Journal* forecast (November 20, 1975), meeting the enrichment needs of a nuclear economy by building twelve new gaseous enrichment plants would cost more than $60 billion between now and 2000.

for power transmission lines (more than seven million acres have already been set aside for this purpose.)[29]

Another major environmental benefit would be the opportunity to avoid any further unnecessary additions to the heat burden of the earth. Scientists have found that the total heat from fossil fuel burning or nuclear fission on earth, and the effects of air pollution, may trigger major climatic changes on the globe. Although these changes are not imminent, they would be fearsome, consisting either in sudden cooling and the onset of an ice age, or in sudden warming, polar ice cap melting, and large-scale flooding.

Finally, extracting less energy from the earth husbands the earth's rare, precious stock of fossil fuel molecules, which took hundreds of millions of years to form and accumulate. The complex hydrocarbon chains in petroleum are necessary for many medicines, plastics, fertilizers, and other petrochemicals. One day, future generations may regard our current energy extravagance as not only thoughtless, but antisocial and even criminal.

In addition to the environmental benefits of energy conservation, conservation offers some economic advantages. First, it saves us money; secondly, it tends to exert a downward pressure on fuel prices and, to the extent that this lowers spending on energy, it weakens the energy combines. But the reduction in fuel prices does create a dilemma for conservationists in that the lowered fuel prices will tend to raise energy demand. However, the new equilibrium demand will usually still be less than the preconservation demand level.

As fuel demand falls due to conservation, it becomes unnecessary to mine or drill less accessible and hence more expensive fuel deposits. Consequently, the average cost of fuels tends to drop, or at least to rise less rapidly. And because energy is an intermediate good used in producing almost everything, holding fuel costs down tends to slow the rate of inflation throughout the whole economy.

Energy conservation also means that jobs are created. This occurs because the energy industry is highly mechanized with high capital requirements and low labor requirements. As energy conservation begins taking effect, consumers find they have to pay less and less of their incomes directly to the energy industries. The money saved can then be spent in service or manufacturing industries that produce far more jobs than energy industries.

Finally, because the energy industries use the largest amount of investment capital of any industrial group in the U.S., and because the capital demands of this sector have been rising steeply in the last decade, their unrestrained growth can deprive other industries of needed capital and can push up interest rates. Without energy conservation, soaring energy demand will lead the energy industry to invest in ever more expensive energy sources, draining the nation's wealth into the energy sector, to the detriment of other branches of the economy. Once this capital is committed to energy production, large amounts of it will be easily siphoned off into the huge financial institutions that control energy in the U.S. (See Chapter 8)

DECENTRALIZED ENERGY

Far more extensive energy savings than any discussed so far would be possible if a remarkable decentralized energy system proposed by Department of Transportation consultant Frederick M. Varney meets its developers' expectation and is widely introduced. Varney's system, still in an early stage of development, is only one of many ways conservation may eventually be implemented. As we have already described the more mundane and immediately applicable conservation approaches using more energy-efficient buildings and more efficient transportation, let's now consider Varney's more advanced and speculative proposals. His system is based on the use of high temperature heat energy; a highly efficient *external* combustion engine; a conventional heat pump; and newly developed high-energy-storage salt mixtures. Used in conjunction with each other, these devices are virtually non-polluting, and they can be adapted not only to buildings and shopping centers, but to factories, farms, and even cities. The system, according to Varney, will make possible a drop of approximately 90 percent in fuel required for household energy requirements. Comparable fuel savings would occur elsewhere in society. Moreover, these savings would enable home-owners to disconnect from big utility grids while having all the power they needed.

The system relies either on the burning of a small amount of low-grade fuel or on the use of an alternative energy source, such as the sun, to raise heat for the operation of a highly efficient closed gas-cycle engine, of which the Stirling-cycle engine has been the most intensively developed.

In a closed gas-cycle engine, fuel is not burned in the engine's cylinders, as in an internal combustion engine. Instead, combustion occurs constantly in a separate chamber. A gas, such as hydrogen, is then circulated through a heat exchanger that warms it and causes it to expand. The energy of expansion is transferred to the engine drive shaft by means of a piston or turbine. (In a Stirling-equipped Ford Torino currently being tested by the Ford Motor Company, the energy of the pistons causes a "swash" or wobble plate to rotate, turning a drive shaft.) The gas then gives up much of its remaining heat through a cooler, like a radiator, and the heat is transferred in a regenerator to preheat that portion of the gas about to be expanded. Because combustion is constant rather than intermittent, as in spark-ignited engines, the temperature of combustion and the fuel/air mix can be well controlled. This results in much more complete combustion and greatly reduced pollution.

The Stirling has a series of impressive advantages over internal combustion engines:

(1) It can run on stored heat instead of fuel, making it compatible with solar, geothermal, and wind-operated heat sources.

(2) It obtains excellent mileage, and can cleanly burn a wide variety of fuels including gasoline, heating oil, methane, methanol, hydrogen, coal, and wood, as well as various low-grade, poorly refined fuels.

(3) It is so much more durable than internal combustion engines that its life-cycle costs are likely to be far lower. Engine lifetimes of two hundred thousand miles for passenger cars and a million miles for trucks and buses reportedly will be achieved.

(4) It produces *no* air pollution when powered by heat, and exceedingly low pollution when combusting hydrocarbons. With the addition of emission-control devices, its pollution can probably be reduced to near zero.

(5) It is almost silent when operating.

In a decentralized energy system, this superior engine is used (1) to power an efficient heat pump to provide heating and cooling of buildings, and (2) to generate current from an alternator to power electric appliances. Surplus heat from the heat pump and heat from a high-temperature solar collector are stored in a fluoride salt mixture. This rechargeable heat store could remain at temperatures of 1000°F-1500°F for periods of weeks. Heat from this thermal battery would be available on demand to power the Stirling engine and to charge a smaller thermal battery of a specially designed automobile, powered by another Stirling.

The Varney system theoretically is compatible with a variety of heat and power sources to operate the Stirling engine. It could use heat from a high temperature solar concentrator or from an available geothermal well. It could also be connected to a windmill-powered heat pump. It should be emphasized that use of the Stirling in a decentralized energy system does not depend on the achievement of any new scientific breakthroughs: Stirlings have been around since the early 1800s and can run with great efficiency when stoked with conventional fuels. A $500,000 study recently completed by the Jet Propulsion Laboratory for the Ford Motor Company found that the Stirling could be perfected by 1985 with 30 to 45 percent better mileage than internal combustion engines.[30]

The fuel savings possible in decentralized energy systems based on the Stirling depend on the source of primary energy used. The Varney system using fuel combustion for primary heat requires a third or less than the amount of fuel needed to generate equivalent power from a central station source. If, instead, the system uses the sun as its primary source of energy (focused with a high temperature solar concentrator) but with a standby source of hydrocarbon fuel, Varney claims fuel savings of 90 percent could result.[31] Obviously, if the system relies entirely on the sun for its primary heat, it requires no fuel whatsoever.

Although the Stirling engine has been used commercially for many years, only in the last three years has it been made small, light, and efficient enough to be put in cars. Test results so far indicate emissions are far less than Federal emission standards for 1977, and that if the Stirling is equipped with emission control devices, the engine's pollution will practically disappear. The Stirling also wins high marks for durability: one model endured a bench test roughly equivalent to a million miles of driving at a hundred miles per hour, with almost no wear.

Varney proposes using the Stirling in conjunction with a heat storage battery. The storage medium is a fluoride salt, such as lithium fluoride, magnesium fluoride, or sodium fluoride. These salts are inert and nontoxic. They can absorb heat from the combustion of any fuel and from non-combustion sources such as solar heat concentrators, geothermal sources, or windmill-operated heat pumps. The salts are able to hold heat energy for weeks, providing power while the sun is not shining or while the wind is still. These batteries are able to store a megawatt-hour per cubic meter, and could be used to run Stirling engines in heat-powered automobiles. In the not-so-distant future, our highways may be traveled by pollution-free, fuelless, Stirling-engine cars!

Another major advantage of the fluoride heat store is its potential as an electrical load-leveling device. When central power station demand for electricity is low, current could be drawn for heat pumps to charge heat stores. During periods of high power demands, these decentralized energy storage vaults would "pay-out" heat energy for conversion to useful work. According to Varney, by such load leveling, twice the generating and distribution capacity of a central system then becomes possible. Because, as explained earlier, a heat pump sops up and bails out energy from a heat sink, such as the ambient atmosphere, into a collector (fluoride salts) at very high efficiencies (except in very cold weather), use of heat pumps throughout a power-consuming system exerts a multiplier effect that concentrates two or more times the energy required to run the heat pumps in the collector. The load leveling and the multiplier effect mean that far less central station plant is necessary, except in extremely cold weather.

The superior efficiency of power from a Stirling-driven alternator over central power generation implies that, in an energy-scarce world, we should get more electricity per unit of fuel from decentralized generation based on Stirlings than by building new power plants.

Varney's research indicates that Stirling-driven generators use two-thirds to four-fifths less fuel than do central stations and that Stirling-driven heat pumps assisted with flat solar heat collectors use only 12.5 to 20 percent as much fuel as either oil or gas furnaces. Finally, Varney estimates that the relative dollar costs of building decentralized energy systems are vastly less than adding

11

DAZZLING ENERGY
Direct from the Sun

Anything that slows down the development of solar energy — the one cheap, limitless source of energy that cannot be shut down by war or embargo — is undermining the national security.

— *The Role of Small Business in Solar Energy Research, Development, and Demonstration,* Interim Report of the Select Committee on Small Business, U.S. Senate, October 9, 1975

The fossil/nuclear industry is fighting the rapid development of solar electric power which poses such a serious threat to the continued economic and political power of that industry. This explains, in part, why our company and another small company, International Solarthermics, have suffered intimidation and harassment. This explains, in part, why Exxon, Mobil and Shell have bought out small solar electric power companies. This explains, in part, the government's long time-frame for the development on a commercial scale of solar electric power. This explains, in part, why the larger, more powerful companies, especially those which have the most to lose, are getting many of the federal grants in solar energy research and development . . . . I do not, however, believe that the Government can continue to cover-up the sun.

— Statement of Edwin Rothschild, Consumers Solar Electric Power Corporation before Consumer Affairs/Special Advisory Committee on the Federal Energy Administration, April 17, 1975

Halfway between San Francisco and Los Angeles in the town of Atascadero is a plain one-story split-level with a carport and patio. Except for the plastic sheath visible on its roof, the dwelling appears nondescript. Yet this house, "Skytherm Southwest," gets 100 percent of its winter heat from a solar space conditioning system on the roof.[1] The tenants need no auxiliary heat and, in summer, they have found the solar cooling system superior to conventional air conditioning.

Heat for the Skytherm house is collected in four large, water-filled bags of transparent plastic. These hold a total of sixty-three hundred gallons of water, and are covered with a plastic sheet sealed to the edges of the bags. During the day in winter, this sheet is inflated to help the bags trap solar heat, which radiates downward into the house through the ceiling. On winter nights, the bags are covered with sliding panels, to minimize heat losses. The panels are moved automatically by a quarter-horsepower motor. The same rooftop "heat reservoir" cools the house in summer: the air bags are covered during summer days to reduce their heat uptake, and they are uncovered at night to radiate heat absorbed from the interior of the house to the night sky. An expert for the Housing and Urban Development Administration estimated that, excluding land costs, the house could be built for $27,500. The Skytherm design is known as a passive solar heating system — one not requiring complex pumps and valves or other mechanical devices to circulate heat, but relying instead on natural heat convection.

A fundamentally different solar heating system is being installed in the north campus building of the three-hundred-thousand-square-foot Community College of Denver, thanks to a grant from the Colorado state legislature. The college will use solar collectors on its roof to trap solar heat in a water-ethylene glycol mixture (to inhibit freezing). The absorbed heat will be transferred to two huge underground water storage tanks, and a heat pump will send it from there to heating-coil-and-fan room heating units in the winter, and to a cooling tower in summer. The system is designed to deliver 100 percent of the building's heat.

Various other buildings completed or in progress use solar heating systems coupled with solar cooling. At the Towns Elementary

School in Atlanta, Georgia, the school is expected to get 60 percent of its cooling from a hundred-ton ARKLA lithium bromide absorption chiller, run mainly with water heated to about 200°F by rooftop solar collectors.

Although there are as yet only two hundred solar-heated or cooled buildings in the U.S., sun power is gaining adherents so fast that within two or three years, thousands of solar buildings will glint in the sun across the country. The efficient, energy-conserving performance of solar buildings has virtually insured their popularity.

SOLAR IN PERSPECTIVE

The development of solar power in the U.S. is closely linked to the nation's nuclear commitment. Because the giant energy corporations know that solar power development could make further nuclear fission plants unnecessary, they are deeply opposed to it — until their gargantuan investments in the nuclear boondoggle can be recovered. Only after trying to exact this economic tribute from society will major corporations begin large-scale commitments to solar power. Present token solar efforts by several large corporations are designed more to capture lucrative government grants than to pioneer the solar industry or promote the use of cheap solar energy devices.

While the U.S. for thirty years struggled to dispel the stigma of being first in the world to develop and use atomic bombs, U.S. scientists sought almost feverishly for peaceful uses of nuclear energy. Their pursuit has been lavishly supported with federal funds, perhaps to ease our national guilt about the bomb. Five billion dollars in government money and scores of billion in industrial capital have fueled the efforts to produce safe energy from the atom. Now a quarter of a century of experimentation with the "peaceful atom" has passed, and significant results are in: a hazardous nuclear industry stands poised on the brink of economic collapse, pleading for more billions in public funds to prevent its demise.

Meanwhile, the orgy of profiteering on nuclear fission power has robbed taxpayers of the legitimate benefits their research and development dollars could have provided. The U.S. *could* have made a real contribution to world welfare by pursuing a vigorous solar energy development program in the early fifties, instead of chasing a will-o-the-wisp, "safe nuclear fission."

Although the development of solar power was advocated by scientific advisers to President Harry Truman, their advice was largely ignored. In the Paley Commission report, those presidential advisers said, "Efforts made to date to harness solar energy economically are infinitesimal. It is time for aggressive research in the whole field of solar energy — an effort in which the United States could make an immense contribution to the welfare of the free world."[2] That was in 1952.

In the eighteen years after the report, solar energy research and development received only an average of $100,000 per year in federal support. During much of this time, while large energy conglomerates were thriving on fat contracts for nuclear weaponry and commercial reactors, solar energy was misrepresented as an "exotic," long-range energy option that held little immediate potential as an energy source. Today the U.S. spends $25 billion a year to import fully 38 percent of its oil supplies.[3] Had solar received support comparable to nuclear fission's, a substantial part of these huge sums would now be unnecessary.

Solar energy can still provide most of our energy needs and liberate us from dependence on nuclear power and foreign energy sources, provided only that the nation makes the *political* decision to develop it rapidly. The major obstacles are emphatically *not* scientific ones. A basic problem is the lack of awareness, even among scientists, of the tremendous recent advances in solar technology. With the information in this chapter readers will be more aware of these developments than are a great many scientists and government energy policymakers. This is not to suggest that the main reason for our slowness at using solar energy is a pervasive naivite on the part of people who ought to know better. A number of corporations with large nuclear commitments have belittled solar power in an effort to buy time for the redemption of their unwise nuclear investments. And federal energy policy has furthered their ends, in general.

SOL, ITS NATURE AND ADVANTAGES

The sun is actually a gigantic, lustrous fusion reactor, far bigger, hotter, and — at a distance of ninety-three million miles from earth — most probably safer than any we are ever going to build on earth. It deluges the planet with an enormous energy field, far beyond our energy needs. While the energy companies talk of energy crises, the surface area of the U.S. alone receives about nine thousand trillion kilowatt hours per year in solar energy, an amount roughly six hundred times our current energy use. An average of 0.7 kilowatts falls on every square meter of land during daylight — about 64 watts per square foot. Even in the rainy northwestern U.S., the equivalent of three times the average household electricity use can be collected from the rooftops of those same residences.[4] And viewed from a global perspective, the solar energy reaching the Saudi Arabian desert each year equals the world's entire reserves of oil, gas, and coal.[5]

The earth's daily solar energy bath not only heats our planet, but provides the means for all plant growth, for all weather phenomena, and for powering the oceans' currents. So wind energy, hydroelectric power, and fossil fuel energy are all fundamentally just different kinds of solar energy.

The winds occur as a result of the expansion and contraction of the atmosphere caused by solar heating. Hydroelectric power from flowing water is made possible by sun-powered water evaporation and subsequent rain. Fossil fuels are but the remains of prehistoric plants and animals, transformed by physical and chemical processes into the familiar hydrocarbons we know as oil, coal, and gas. Direct solar energy, in the form of heat and incident sunlight, supplies us with fuel, warmth, coolness, food, and fresh water. The versatile sun also directly can supply us with temperatures hot enough to propel vapor-driven power plants and with refrigeration.

The technology for using the sun is safe and can be used with minimal damage to the environment or the public. No energy technology is without *any* adverse effects, and solar power is no exception. Some solar power plants will add to the heat burden of the earth by increasing the absorption of solar energy and then converting it at relatively low efficiency to electricity, with resultant releases of heat. This is not a serious obstacle to the use of solar power and thermal pollution is common to all power technologies.

Unlike fossil or fissile energy, the energy from the sun is virtually inexhaustible and can be tapped without exhausting limited energy resources, save for those used in constructing and maintaining the solar plant. And because no fuels are burned in the course of plant operation, solar power plants will not foul the air with particulate or radioactive discharges. Thus divorced from fossil and fission fuels, solar power has an inherent simplicity and elegance that no other technology can match. Another great asset of solar power is that it is available from the sun in large amounts during the late afternoon — when the greatest electrical demand occurs. Moreover, a large proportion of the solar energy needed is distributed free by nature to the point of use. We do not need to pay an energy conglomerate to bring the sunshine to us, unless we let those corporations "pull a fast one" on us as they have done with nuclear fission.

Using solar energy effectively is especially challenging because the sun is a diffuse and an intermittent energy source that varies in intensity daily, seasonally, and by geographic region. These characteristics add to the complexity of solar systems and hence to their costs. But the problems are being successfully overcome. One reason for this success is that, in contrast to nuclear fission or fusion, the technology required to utilize most forms of solar energy is astonishingly simple and familiar. Solar technology for home heating and cooling has been available for at least twenty years. Today, that technology along with other energy alternatives can provide many citizens with total energy independence from power utilities and relative independence from big energy conglomerates.

Some U.S. energy experts are already willing to acknowledge how close the U.S. is to effective energy decentralization. For example, in 1973, a landmark review of solar energy technology appeared. It had been prepared for the AEC by a multi-agency technical review panel known as Subpanel IX, chaired by Dr. Alfred J. Eggers, Jr. of the National Science Foundation, and it bore the bland title, *Solar Energy Program Report*. But it concluded, in what was then a bold statement, "Ultimately, practical solar energy systems could *easily* contribute 15 to 30 percent of the nation's energy requirements." [6] [emphasis added]. The subpanel also found that solar energy could provide 30 percent of the nation's energy for heating and cooling buildings by the year 2000, and that photovoltaic solar cells could be produced at economically competitive prices by the mid-1980s.

Another report prepared at about the same time was furnished to the National Science Foundation by the Mitre Corporation. In *Solar Energy Research Program Alternatives,* Mitre estimated that solar energy's potential energy contribution could go as high as 35 percent of all U.S. energy in the year 2000. [7] In view of the solar technology advances of the past five years, these once bold projections may actually prove to be conservative, for reasons explored below.

TWO KINDS OF DIRECT SOLAR ENERGY CONVERSION

Solar thermal processes use the heat of the sun directly without converting the sun's energy to electricity. Solar thermal power can be used for warmth, for melting substances, or for running an engine. When used to power refrigeration or air conditioning equipment, it produces cooling. Solar thermal energy's most common use is in heating water and in heating and cooling buildings, but engine operation by solar thermal energy has been well understood for more than a hundred years, and many solar pumps and heat engines have been built from 1870 to the present. As early as 1913-1914, a solar thermal steam engine that developed fifty horsepower was built near Cairo, Egypt, by an American engineer, Frank Shuman. [8]

As distinguished from solar thermal processes, a second major direct use of the sun is in solar electric systems, of which photovoltaic processes are currently the most important. They are based on the known property of certain thin substances such as single crystal silicon, cadmium sulfide, or copper sulfide to produce a current when exposed to light. Solar cells of silicon have been known for twenty years and have powered most American space vehicles. The direct use of solar power to make electricity is the most expensive and technologically complicated aspect of solar power production today.

Solar cells can convert solar energy to electricity either in small solar units on private homes and businesses, or at large centralized power stations. Because solar cells lend themselves to efficient, decentralized use, the solar power economy is not subject to catastrophic power failure as is a nuclear economy. A solar society would be much less vulnerable to the blackouts or brownouts that can occur when large conventional power stations have to be shut down.

SOLAR THERMAL USES

The most practical and important immediate application of solar thermal power is in the heating and cooling of buildings and in the heating of water for residential and commercial use. Rooftop solar water heaters were widely used in the U.S. in the early part of this century, particularly in California, Arizona, and Florida, where they were sold by the tens of thousands in the 1930s.

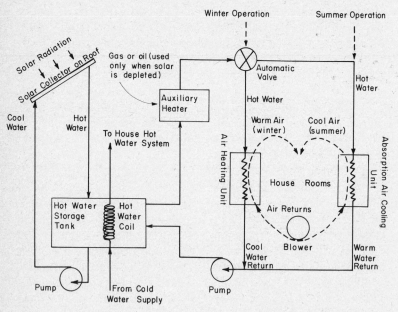

Figure 11-1. Residential Heating and Cooling with Solar Energy.

The basic element of a conventional solar hot water system and a solar heating and cooling system is a flat, thin, rectangular heat collector box containing a flat-plate collector to absorb the sun's heat. The collector is usually made of sheet metal, such as aluminum or copper, and is blackened with a coating to increase incident heat absorption. To prevent sunlight from being reradiated as heat from the metal plate back into the sky, the collector is covered with one or more transparent layers of glass or plastic to trap infrared radiation. The entire box is sealed and connected at one end to an inlet pipe and, at the other, to an outlet pipe so that water can be circulated through the box. By tilting the box at an angle with brackets and placing the inlet pipe at the lower end, water can be induced to rise by convection over the plate to absorb its heat. For efficient operation, the collector surface may be corrugated to distribute the water flow evenly, or it may have tubing welded onto or built into the plate surface so water can flow through. In some designs, water is pumped to the top of the collector and allowed to trickle down slowly and evenly over the collector.

Less commonly, flat-plate designs may use circulating air to transfer the heat stored in the collector. If water is the heat-transfer medium, it is piped off into an insulated storage tank for later use. If air is used, it may be used to heat a storage tank, usually containing water, gravel, or special salts. This tank is connected to a system of water or hot-air radiators and to a pump so that when heat is required in the house, a thermostat activates the pump (or fan) to circulate heated water (or heated air) from the heat store throughout the building.

For optimum efficiency solar collectors should be tilted towards the sun at a particular angle, depending upon the latitude of collection. Simple in design and easily maintained, domestic solar water heaters were once popular and in wide use, until cheap natural gas was introduced along with mass-produced fossil-fired water heaters. Today, because of the upward spurt in fuel prices, solar water heaters have again become economical. They are easy to build or can be purchased and installed by contractors for about five hundred dollars. They have been mass-produced in Japan, Israel, and Australia, where they have sold for about two hundred dollars. Their widespread introduction can result in important fossil fuel savings as about 3 percent of U.S. energy use today is for hot water heating.[9] Affluent Americans also are starting to use specially designed units to heat swimming pools. Commercial applications of solar water heat are straightforward and similar to domestic applications.

Solar water heat also has industrial applications. The Sohio Petroleum Co. of Cleveland, Ohio, is considering the installation of a dozen two-hundred-foot long solar ponds to heat water for its uranium extraction plant at Grants, New Mexico. The ponds are expected to raise water from wellhead temperatures of 60°F to 130-140°F in the summer, and to 80-90°F in winter. The water is stored in underground reservoirs to minimize heat losses until the water is needed. This should lower Sohio's fuel oil needs by twenty thousand barrels a year. Preliminary feasibility of the idea has already been demonstrated by the Lawrence Laboratory of Livermore, California. If successful, this solar heating process may be widely used throughout the western U.S. where both uranium and copper mining industries rely on heated water to leach their ores.

As noted, solar space heating employs essentially the same technology as solar water heating. The flat-plate collector is used and heat from it is drained into an insulated storage tank. Like solar water heating, home solar space heating is in commercial and residential use today. It is already economically preferable to electric heating and, in some areas of the country, to oil and gas heating. Moreover, its competitive position is steadily increasing as oil prices increase, and as gas grows scarcer. Considerable latitude exists in the estimates made by various experts on the costs of solar heat, and the payback period needed to recover first costs. Although the commercial first cost of solar heating systems is relatively high, three to eight thousand dollars (at least 10 percent of the home's cost), operation of most systems is cheap, and the first cost in many cases can be recovered with interest during the system's life. Expert sources of the *New York Times* however, place the commercial cost of installed solar heat at eight to ten thousand dollars, on the average.[10]

For most residential and commercial solar projects today, solar heat is a supplemental rather than a total source of heat. Although there are no technical barriers to building a 100 percent

Wescott/ERDA

Figure 11-2. Solar panels at the Timonium Elementary School near Baltimore, Maryland. The panels collect the sun's heat to power a solar heating and cooling system for the school.

solar heating system, solar heat tends to become uneconomic when efforts are made to provide more than about two-thirds of a building's heat.

This drawback has been conquered in a number of instances by personal effort and ingenuity. Robert and Eileen Reines, who live near Albuquerque, New Mexico, created a dome that is totally heated by solar power, and is totally electrified by windmills. Resembling a spacecraft amidst barren rocky surroundings, the dome's hemispheric shape and various other features reduce its heating needs so the Reines stay comfortable even in sub-zero temperatures during winter and keep from sweltering during the hot New Mexico summers. The house cost $12,500 to build, including the solar system and the wind generators! Inside the dome are all the amenities, including stereo, TV, radio, power tools, and an electric toilet. Only the stove runs on butane.

When a conventional solar home is sold, the heating system constitutes an additional equity, similar to the value of an extra room or patio. Homeowners' resistance to incurring the high first cost of installing solar space conditioning could be greatly reduced if the federal government allowed a percentage of investments in solar energy systems to be deducted from tax bills — a type of "reverse depletion allowance" for conserving scarce fossil fuels, as one solar manufacturer has suggested. A bill introduced recently by Senator Gary W. Hart and others would provide low-interest loans to builders and owners of solar units. Another bill by the same legislators would require energy conservation in government-owned-or-financed buildings. Four hundred thousand buildings are owned and operated by the government so this legislation would spur solar development greatly.

Dr. Jerry Plunkett, President of Materials Consultants, Inc., a small solar heating company, has estimated that solar collector cost — the largest item in solar heating equipment — could be reduced by a factor of ten through mass production.[11] This will result in economies of scale, and construction of collectors as integral roof components will lead to additional savings in building construction costs.

The Los Alamos Scientific Laboratory already has developed a flat-plate collector that serves as part of the roof and insulates the building. Flat steel plates are spot-welded and pressure-expanded to form heat transfer panels. Glazing is placed over the plates and insulation is put beneath them. Los Alamos forecast that the plates can be built for $2.50 to $3.50 per square foot. Thus a four-hundred-square-foot collector would cost twelve hundred dollars and replaces some roofing material. Within ten years, Dr. Plunkett forecasts, home solar heating, including improved insulation and storm windows, can be available for a total of twenty-five hundred dollars.

Because solar heating systems are by no means limited to tropical or subtropical areas, such as the southwestern desert of the U.S., their use in areas of high fossil-fuel costs, such as New England, already can have major impacts on homeowners' fuel bills. In New Hampshire, where a $1,000 annual heating bill is not uncommon, installation of a typical solar heating system today would result in significant annual fuel savings.

Consequently, residential use of solar heat is increasing rapidly today despite its sizable first costs and solar energy researcher Dr. George O. G. Löf expects thousands of homes to benefit from solar heat systems within a few years. Commercial use of solar heat is becoming more common, too. The Solaron Corporation of Denver has recently installed solar heat in a forty-one-thousand-square-foot, million-dollar commercial headquarters for Gump Glass Co. in Denver. The system is expected to save 77 percent of the annual fuel costs. [12]

Although space cooling is not as close to widespread commercial use as space heating, it, too, is becoming more practical. Dr. Erich Farber, Director of the University of Florida's Solar Energy and Energy Conversion Laboratory, has developed prototype ammonia and water refrigeration systems in which solar heat is used to drive volatile ammonia gas from solution in water. The gas is collected, condensed, and subsequently evaporated, all in a closed system, to cool its surroundings. Another cooling system uses solar heat to operate mechanical compressors to power conventional refrigeration units. This absorption refrigeration process can use a lithium bromide solution [13] as at the Towns Elementary School, described earlier.

Numerous other solar thermal applications are economical and proven. Solar stills that use the sun's heat to evaporate water, leaving dissolved salts behind, have been in use for many years.

The desalinated water can be used for the irrigation of desert regions. There, pumps run by solar energy can be used to distribute the water. Solar heat also has important applications in agriculture for drying crops (during the night as well as by day) and can be used to heat greenhouses and aquaticulture ponds.

Solar stoves, ovens, kilns, and furnaces have also been operated successfully. Like most heating and cooling solar technology, with the exception of photovoltaic cells, stoves and ovens can be built by the do-it-yourselfer.

SOLAR PHOTOVOLTAIC POWER

Instead of using the sun's heat directly or transforming it with mechanical energy, photovoltaic substances turn sunlight directly into electricity without using moving parts. Photovoltaic materials, fabricated into discrete units with positive and negative terminals, form solar cells. When a photon of light hits these thin semiconductors, the photon creates a negative and positive charge known as an electron hole pair. Ordinarily, the charged pair would recombine into a neutral charge in a millionth of a second, but because of a junction between two regions of the silicon semiconductor, the tiny charges are segregated and an electrical potential is generated. A current then can be drawn off by an external circuit. Cells can be connected in series and parallel with others in an array to generate large amounts of electricity. Arrays of cells packaged in supporting frames form a solar panel.

Solar cells of selenium have been known for about a hundred years, but their efficiencies were very low. A much more efficient cell was produced by Bell Laboratories in 1954. This cell had another great advantage: it was made almost entirely of silicon, the second most abundant material on earth. Silicon forms a quarter of the earth's crust and can be cheaply obtained from sand. A high degree of purity in the silicon is required, however, so the silicon must be well refined. Then the pure material is "doped" with special impurities that give it its photovoltaic properties. Although solar cells can be made from cadmium sulfide, copper sulfide, and gallium arsenide, as well as other compounds, silicon is the most frequently used.

Solar cells are simple in operation and potentially extremely cheap to manufacture. Because no material within the cell is consumed by its operation, they can last indefinitely if made properly and the cell junction protected. In practice, although the silicon cells themselves are quite durable and withstand high temperatures well, cell packaging materials, junctions, and electrodes are subject to deterioration. As the technology of solar cell manufacture continues its rapid advances of the past twenty years, however, the wear-resistance of cell arrays will undoubtedly increase.

Cell arrays have been used to power space satellites and harbor lights, walkie talkies, light meters, and microwave relay stations. Often these applications today are in remote areas where they are seldom seen, but as soon as cell costs fall sufficiently, cell arrays will be placed on rooftops, fences, patio covers, garages, and in large-scale central station power plants. The technology to do this is functioning, well developed, and well understood. The only non-institutional barrier to rapid solar electrification of the U.S. is the cost of solar cells.

SOLAR CELL COSTS

Very high estimates are often made for solar cell costs. Solar cell prices are quoted at between one and two hundred thousand dollars per kilowatt for "space-rated" solar cells.[14] Considerable divergence of opinion exists not only on what the current costs of terrestrial solar cells are, but on the potential for reducing those costs. Nonetheless, numerous experts believe that photovoltaic cells could produce commercially competitive power within ten years. (The immediate cost goal of ERDA in the solar cell field is to produce solar cells at five hundred dollars per peak kilowatt by the mid-1980s.[15]

It is widely accepted that solar cells become competitive for home use at five to seven hundred dollars per peak kilowatt, and at three to four hundred dollars per peak kilowatt for central station power generation. (The peak watt output of a solar cell is the amount of power produced under optimal conditions on a bright sunny day.) Because the plant produces no power at night and reduced power in cloudy weather, about three times the plant capacity must be installed to provide an average kilowatt (on a twenty-four-hour-a-day basis) as contrasted with a peak kilowatt.

Therefore, the average cost per kilowatt of solar plants is about three times its peak kilowatt cost. Even without any electrical storage capacity, such plants can be useful for providing load following or peaking capacity, that is, for meeting fluctuating demand (above a steady base power level) during daylight. Until solar plants with storage capacity come into general use, conventional plants can carry the baseload. Electric storage capacity can be added to solar plants later at an additional cost (but not exceeding twice the original cost), so they can provide power on a twenty-four hour basis. Although the five-hundred-dollar per peak-kilowatt figure is widely regarded as the threshold of economic competitiveness for solar electricity, particularly as fossil fuel and nuclear plants continue their rapid rise in costs, some solar proponents believe solar cell plants can be competitive even at much higher peak-kilowatt costs.

The good news about solar photovoltaic costs officially began arriving in the early seventies in basic studies by the National Science Foundation/National Aeronautics and Space Administration Solar Energy Panel (December 1972),[16] by Subpanel IX (December 1973),[17] and by the Mitre Corporation, in its report to the NSF.[18] "Preliminary estimates indicate," Mitre concluded, "that by 1990 the price of electricity produced by photovoltaic systems may be reduced to about 10 to 15 mills (1 - 1.5 cents) per kWh, if the cost goals of the system are achieved. This would be competitive with electricity produced by 1990 designs of fossil-fueled turbogenerator plants, burning clean derivatives of coal, i.e., about 13 mills per kWh."

Even more favorable projections were made by a consultant to the NSF/NASA study, Dr. Paul Rappaport, director of RCA's Process and Applied Materials Research Laboratory. Testifying in Congress in June 1974,[19] Dr. Rappaport stated that five hundred dollars per peak kilowatt could be achieved by 1985 and that one hundred dollars per peak kilowatt would be attainable by 2000AD if cell efficiencies ranged between 18 and 20 percent, and fifty thousand peak MW of capacity were produced annually. Estimates by another participant in the same hearings, Dr. H. Guyford Stever, director of the National Science Foundation, were consistent with Dr. Rappaport's conclusions.

Since the NSF/NASA and Subpanel IX reports were published,

additional information on solar costs has become generally available indicating that ERDA's current estimates of solar photovoltaic power's potential are conservative. The ERDA estimates state that solar power *of all kinds* could contribute only 7 percent of our energy in 2000AD.[20] This happens to be exactly the electrical contribution that Subpanel IX said could be made by photovoltaics *alone.* ERDA's unrealistically low estimates relegate photovoltaics to producing a minuscule amount of our energy by the year 2000. The true situation is much brighter.

As the facts presented by Drs. Rappaport and Stever imply, in less than ten years solar cell power can be competitive with fossil-fuel power on a peak-load basis. Solar cell power then can be substituted for peak-load fossil fuel power, liberating fossil fuels to supply baseload power, and making it unnecessary for us to resort to uranium fission for the baseload. Furthermore, once solar energy storage systems are perfected, the use of fossil fuels, even for baseload power, will become less necessary.

Solar cell cost estimates today are highly dependent on the total volume of U.S. cell production. Mass production and automation are the keys to lowering solar cell costs, in conjunction with the development of new cell materials and fabrication processes. Cadmium sulfide cells, for example, can be made more cheaply than silicon cells, but their lower efficiency is still a problem. The production of such solar cells by the deposition and evaporation of thin semiconductor films is currently being studied. As cell production volume increases two-fold, costs are said to drop by 25 percent, according to Dr. Joseph Lindmayer. Today, Lindmayer's firm, Solarex Corporation of Rockville, Md. can produce complete solar arrays for seventeen thousand dollars per peak kilowatt or seventeen dollars per peak watt.

Solar cell experts regard even greater cost reductions as feasible in the next five to ten years for the whole solar cell industry. A team of scientists from the Jet Propulsion Laboratory of Pasadena, California, analyzed a solar cell mass production technique proposed by Mobil Oil Co. in affiliation with Tyco Laboratory in Waltham, Massachusetts. JPL concluded in 1972 that Tyco's process could ultimately produce silicon solar cells for $250/peak kW, and that large solar cell power plants could be built for $375/peak kW.[21] This would make them not just competitive with fossil plants, but *strongly preferable* to coal plants of $600-700/kW or more. Finally, as the inevitable depletion of finite fossil-fuel reserves continues, solar photovoltaic plants will become progressively more and more preferable as they require no fuel, create no pollution, can be made immune to catastrophic failures, and have low operating costs.

The primary cost element in silicon cell manufacture today is labor, not raw material. Pure silicon costs less than thirty dollars per pound, and in cells of .004 inch thick, a pound goes a long way. Further reductions in the cost of silicon are likely soon. The cost has already dropped 13.5 times from four hundred dollars per pound as recently as 1961.[22] The cost of pure silicon will not be a barrier to bringing about a radical reduction in silicon cell costs.

Large reductions in cell cost depend on large production volumes that will make automation profitable. Most cells are still largely handmade. Every time a cell is sliced and ground from a silicon boule or ingot into a wafer, the process now wastes three times as much silicon as each cell contains. Solar cell scientists realized years ago that if a mass production process could be developed to grow silicon crystals of the desired size and shape directly, without the costly, laborious, cutting-and-polishing stage, cell array costs could be slashed.

Tyco has already used its new process to produce one-inch wide ribbons of silicon six feet in length, suitable for use in solar cells. Tyco's Executive Vice President, Dr. A. I. Mlavsky has estimated that, exclusive of packaging, ribbon-type silicon solar cells could be produced by the company at a manufacturing cost of $165-200/peak kW.*

Tyco estimates that the $165-200/peak kW production cost

---

*For his estimate, Mlavsky made the following assumptions:
1) A modest output of 20 MW of cells annually. (That is, 2 percent of a large nuclear power plant's capacity rating.)
2) A modest cell efficiency of 10 percent (much higher efficiencies have been produced.)
3) Modest engineering advances in ribbon-making machines.
4) $50 -100 million in investment over four years (a scant expense compared to the expected $10 billion cost of breeder development.)
5) And substantial reductions in silicon cost. ("It is predicted," Mlavsky wrote, "that silicon for this purpose can be made to sell at about $10/lb. . . . ")

could result in a cell sale price of twice that amount ($330-400/peak kW). Naturally, scaling up production to produce solar cell arrays by the tens of thousands of megawatts would result in great savings, as Dr. Lindmayer's calculations have shown; Lindmayer forecast that packaged solar panels could sell for $450/peak kW when 1 gigawatt (1000 MW) of cell power are produced annually, while the solar cells themselves would be selling for $225/peak kW.[23] With cell production output increasing by only a factor of 3.2 each year, solar panels could be economically competitive in eight years. This projection is consistent with the Tyco time-implementation projection.

New developments in photovoltaic cells could cause even the cost projections of Mlavsky and others to be undercut. Varian Associates of Palo Alto, California in 1975 announced the development of gallium arsenide disc cells with relatively high efficiency. This cell reportedly can produce 1000 times as much electricity as a silicon cell of equivalent size.[24] Although the compound is more costly than the omnipresent silicon and contains arsenic, it can endure heat that would destroy silicon cells. Varian already has built a rooftop gallium arsenide array of 100 cells that can generate a kilowatt of electricity.

Just as the cost of cells depends on production volume, so it depends heavily on political decisions and government priorities towards solar and other energy technologies. For example, government support for nuclear fission technology today includes the federal manufacturing and provision of enriched uranium to utilities at cost. If the U.S. government decided to promote solar power intensively, it could manufacture solar cells, too, at the least possible cost, and furnish them to users.* Individuals who needed rooftop collector panels and newly established public solar power corporations seeking to build solar plants could all then obtain cells at costs that should resemble the Mlavsky projection of $165/peak kW. That is not bad compared to the $1000/kW costs for nuclear plants being built today.

The use of mirrors, lens, and reflectors has been explored by a number of researchers in an effort to lower cell costs by boosting

---

*Such users could include individuals who wanted their own rooftop collectors for domestic, commercial, or industrial power, and to municipal electric utilities, electric cooperatives, and even investor-owned utilities, provided the public had an alternative to exclusive reliance upon them.

the efficiency of arrays. An experimental solar cell using mirrors to concentrate sunlight has been produced by Dr. Eugene Ralph of the Heliotek-Spectrolab Division of Textron Corporation. The result was a power output 125 times that of the unaided cell. This approach, according to Dr. Ralph, holds the potential for reducing solar cell cost from $15,000/peak kW to $150/peak kW.

While photovoltaics are being perfected, the capital requirements of this industry are so low — compared to the costs of fission technology and fusion development — that great progress could simultaneously be taking place in solar thermal conversion technology, bringing it closer to commercial implementation.

COMBINING SOLAR THERMAL POWER AND SOLAR CELLS

Future development of solar cells is likely to combine solar thermal and solar electric devices so that both electric power and heat will be generated by the same unit. This can have valuable applications for homes and apartment houses where rooftop solar cells could produce current to power appliances, and eventually an electric car. Any excess power generated could simply be fed back into the utility electric power systems, lowering the net electricity taken from the power grid and, therefore, reducing consumers' utility bills. This would add generating capacity to the whole system and would enable utilities to build fewer power plants.

By coating the backs of solar electric cells with selective heat-absorbing substances, heat can be gathered for use in water or space heating and cooling. Dr. K. W. Boer, director of the Energy Conversion Institute of the University of Delaware, has already installed an integrated photovoltaic and solar thermal system on the roof of an experimental house known as Solar One. Air is circulated over the surfaces of the cells and over additional black-coated heat absorbing surfaces. The air then heats a salt heat-storage mixture from which heat can later be retrieved.

LARGE-SCALE SOLAR THERMAL PLANTS

Many physicists and energy experts are not yet aware that the U.S. soon could be getting large amounts of electricity from cost-competitive solar-thermal electric power plants. Practical technology is available *now* to utilize the sun's energy in these plants, and at a lower cost than nuclear power plants and with simpler, locally made building materials. Furthermore, the construction of these plants is relatively labor-intensive and advocates of the technology believe it will produce two to three times the number of "employment years" as would the construction of a comparable nuclear plant.

Great credit is due to two scientists, Dr. Aden and Marjorie Meinel, for calling attention to the necessity of using solar power for the production of electricity. They emphasized the need for practical low or medium technology with which to utilize solar energy immediately, and they did so years before the rest of the scientific community was generally aware of solar power's possibilities. Until recently, the Meinel plant design was the most advanced known. The Meinels estimated that a thousand 1000 MW plants of their design could be built in largely uninhabited areas of the California-Nevada-Arizona desert between Mexico and Las Vegas, Nevada, at an acceptable cost.

Dr. Aden Meinel, a former science adviser to the Secretary of the Air Force, had directed the University of Arizona's Optical Sciences Center. Marjorie Meinel, holder of an M.S. in astronomy, has published scientific papers with her husband and edited scientific journals on nuclear energy and rocketry at California Institute of Technology.

The Meinels projected that their "National Solar Power Facility" eventually could supply most of the U.S. electrical needs and those of Northern Mexico. The plant design they proposed uses mirrors to focus sun on stainless steel or glass ceramic tubes filled with a circulating fluid as the heat transfer medium. The heat is conveyed through a heat exchanger to a steam turbine generator such as those in conventional power plants. Unneeded heat is stored in eutectic salt mixtures (mainly sodium nitrate). Currently, the Meinel design is still in the research and testing stage.

A new solar thermal power plant system, the Smith Multi-module Solar-Electric Plant, has been proposed by Dr. Otto J. M. Smith, a professor of electrical engineering and computer science at the University of California, Berkeley. Smith's design is a great solar advance because it has high thermal efficiency, low land requirements, and needs no large source of cooling water.[25] In addition, major components of the system can be readily obtained at reasonable cost from existing commercial suppliers.

Dr. Smith's design is for a computer-controlled modularized 100 MW(e) power plant. (Several of these plant units could be built to generate much larger amounts of power.) The Smith plant uses

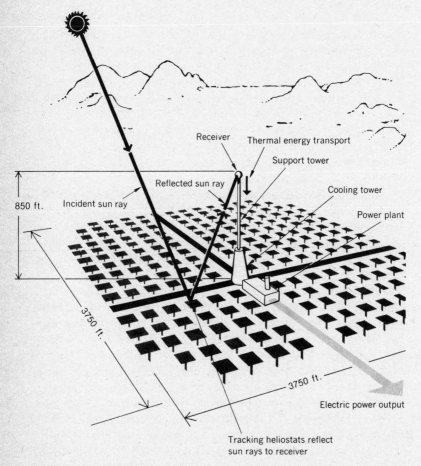

Figure 11-3. Simplified drawing of a 100 MW solar thermal conversion station. Unlike the Smith system, this design uses only one tower for each 100 MW instead of 1,100 towers.

fields of steerable mirrors, which concentrate reflected sunlight on black-coated stainless steel pipes behind heat-conserving windows on short towers. Hot water, steam, and another heat-exchange fluid transfer the absorbed heat from the towers to a central power plant where they are used to operate a conventional turbine and electrical generator.

Each 100 MW plant uses eleven hundred towers about one hundred fifteen feet high and each tower, in turn, is illuminated by three hundred twelve low-cost mirrors. The mirrors are about twenty square feet, and are made of metallized polyester stretched almost flat over lightweight fiberglass-epoxy platens.[26] The entire 100 MW plant takes only three quarters of a square mile in area and has an operating thermal efficiency rating of about 35 percent.* Ten thousand of these 100 MW plants could fit in a square desert area less than 90 miles on a side. Preliminary cost estimates by proponents of this system suggest that each 100 MW unit could be built for $85 million with no energy storage and for $150 million with high-quality, four-hour energy storage, which would enable the plant to meet heavy evening energy demands even after the sun has set.

High solar insolation occurs in Southern California, Nevada, Arizona, New Mexico, and western Texas. These areas are more than adequate for generating *all* of the electrical needs of the U.S. although the transmission of power over great distances is still relatively expensive.

Some scientists, among them H. C. Hottel and J. B. Howard, have been highly skeptical of solar thermal electric power plants, faulting them for alleged low efficiency, high land requirements, and high costs.[28] One reason for their skepticism was their belief that solar concentrators (mirrors) would not be efficient in parts of the country that experience frequent haziness or cloudiness, thereby obliging plants designed for those regions to use low-efficiency flat-plate solar collectors that can capture diffuse sky radiation.

Now this problem seems to have been met by a compound

*For purposes of comparison with fossil- based power generation, ten of these 100 MW solar plants with energy storage, built along the lines of Smith's design, would require only 7.5 square miles — half the land that might be strip-mined to fire a comparable 1,000 MWe coal plant during its thirty-year life, and half the reservoir area needed for some large hydroelectric projects.[27]

parabolic solar collector invented and patented by Dr. Roland Winston of the University of Chicago. The collector concentrates diffuse *as well as* direct solar radiation, and can operate in regions where conventional collectors are useless. The Argonne National Laboratory is building full-scale arrays with the new collector, and the NASA Lewis Research Center is planning to test them.

The Winston collector eliminates the need for a control system for moving the mirrors but in exchange, the Winston device requires trough-shaped fixed position mirrors with much larger reflecting areas. Since solar power plant cost is primarily dependent on reflecting surface area, the Winston design is likely to be more expensive than the adjustable mirror design of Dr. Smith.

ALL ELECTRICITY FROM SUN

Naturally, in a society making enlightened use of its energy resources, many other energy generation sources other than solar thermal power would be available. These sources include wind, geothermal power, some clean coal power, perhaps ocean thermal power, and possibly fusion, all discussed in Chapter 12.

Therefore, it is unnecessary for us to try to get all our electricity from any one technology, such as the solar thermal plant. But if we wanted to produce the entire 6.4 quads of electricity that the U.S. used in 1973 by generating that power from Smith solar thermal electric plants, all the plants necessary could be placed on a square area just fifty-four miles on a side!

By comparison, if we wished to generate all the electricity used in 1973 by means of 20-percent efficient solar cells (assumed to be 10 percent efficient on a twenty-four-hour basis), then only .00176 of U.S. land area would be required — a square of land only seventy-four miles on a side, yet capable of producing all the nation's electricity. With more than 124,000 square miles just in rooftop area in the U.S., it should not be too hard to find anough space to locate all the solar cells and solar concentrators this country needs. As one wag put it, if we run out of space for solar cells, we could try hanging them from power lines. Seriously speaking, anyone who has studied a detailed map of the southwestern U.S. knows how much land is barren and desolate today. Huge areas of this desert lie under military control for questionable use as bombing range and weapons test sites. By repossessing a small fraction of this well-insolated land, which is off limits to most of us now anyway, tremendously valuable solar plants could be built.

SOLAR THERMAL PLANT COSTS

It appears likely from the Smith plant design that solar thermal power is at least reasonably competitive as an energy source right now. As contrasted with the rising fuel costs of uranium fission plants, once the capital investment is made for a solar plant, fuel costs will be essentially zero. Nonetheless, cautious energy analysts may want to reserve judgment on solar thermal costs until actual operating facilities are built. These readers, however, may find it relevant that the existing cost projections for the Smith plant are not far from the solar thermal plant costs listed in the National Science Foundation's Subpanel IX report. The authors of that study found such plants could be competitive with fossil-fired plants by the 1980s, particularly in the Southwest at 25-45 mills/kWh. (The Smith plant can provide power at an estimated 25-30 mills/ kWh.[29] For this reason, it, or other solar thermal plants of different designs, are likely to soon begin replacing large amounts of nuclear capacity.) The Subpanel IX group concluded that solar thermal power could supply a full 30 percent of the country's electrical energy. As explained, in the previous sections we can do even better if we choose.

As with solar cells, the primary barrier to implementation of solar thermal plants is not scientific but political and institutional. ERDA is a major culprit, for its slow-motion approach to solar power development, despite the fact that rapid solar advances could lower utility bills and rapidly create a high-growth, multi-billion dollar solar industry, with numerous jobs for the construction trades and for urban factory workers. But instead of proceeding with dispatch to provide funding for several kinds of 100 MW solar test facilities, ERDA plans to have a 10 MW test facility operating by *1980* and is dallying now with a 1.5 MW test facility at Sandia Laboratories in Albuquerque.

FEDERAL SOLAR POLICY

Evidence of the mistaken priorities used by the U.S. government in allocating federal research and development dollars is reflected in the history of federal energy policy and the pattern of federal spending in the solar energy field.

Although the AEC was well aware of the facts in the NSF/NASA and the Subpanel IX reports, the AEC chose to basically ignore the conclusions and implications of these studies, thereby creating the impression that nuclear fission and breeder reactors were vital. In its first Liquid Metal Fast Breeder Reactor environmental impact statement, the AEC asserted, "The outlook appears to be that solar energy has little potential as an economical major source of electricity for several decades . . . . " Elsewhere the AEC claimed, "little basis exists for projecting a measurable contribution (i.e., by the year 2000) of solar energy to either electricity generation or high-energy fuels since even optimistic projections of cost place solar conversion in a poor competitive position relative to coal or nuclear energy." Yet at the time of the breeder statement, the AEC had the Subpanel IX report! In it, representatives from the NSF, the Department of Defense, the AEC, NASA, and other agencies were saying: "Large-scale photovoltaic systems should be technology-ready by the early 1980s and available in *huge* amounts by 1990," and that solar electric systems could be built for just a hundred dollars per kilowatt (average) in 1990. If solar cells — the most advanced and expensive aspect of solar technology — can become competitive so soon, then clearly, large-scale solar plants suddenly cease to be long-range possibilities and become medium-term options of paramount importance!

But note the follwing inconsistency: although ERDA administrators today are formally conceding that solar energy can be far more important a contributor to our energy needs than the AEC ever contended, ERDA's basic policy — consistent with the AEC's — is still to be gung-ho in extending funding for nuclear fission and breeder reactors, but *penny pinching* in backing solar energy.

Minimizing the solar energy potential is not uniquely a technique of the AEC, of course. General Electric Corporation, Westinghouse Electric Corporation, and TRW Systems Group all made estimates of solar energy potential in 1974 under grants from the NSF (!) and asserted that solar could make only paltry contributions to U.S. energy supplies by 2000 AD (1.6, 3.0 and 5.7 percents, respectively).[30] Numerous other industries allied with the nuclear business and numerous utilities deluged the public with "scientific information" portraying solar energy as an exotic, pie-in-the-sky form of energy that was so expensive we shouldn't even bother thinking about it (we might otherwise have found out how cheap it could really be).* The fears of the big corporations and their research group acolytes are understandable. For if the bulk of the nation's power can be derived from the sun directly, bypassing the big utilities and the nuclear conglomerates behind them, then a wide range of existing energy complex investments may be threatened.

ERDA'S SOLAR FUNDING

Federal funding appropriations for solar research have been stingy and most have been given to a few giant corporations and universities, to the detriment of small but competent solar firms.

Although federal funding for solar energy has increased dramatically from the neglect of the last twenty-five years, only $89 million was budgeted in fiscal year 1976 for federal solar energy research and development — far less than it takes to build even a single, large conventional power plant. As of fiscal 1974, the National Science Foundation was funding 90 percent of all federal solar projects.[32]

Lists released by NSF in 1974 showing to whom 149 contracts and grants in solar energy had been dispensed, indicated that more than half went to schools and fifty-three went to private business and industry. Of the grants to these organizations, $8.7 million — 89 percent of the funding — went to large organizations and only $1.1 million or 11 percent went to small firms.[33] Although ERDA is now charged with the overall coordination of solar energy

*Establishment scientists were no exception to this trend. In the glossy and authoritative-looking volume, *Energy and Power,* consisting of selections from *Scientific American,* Chauncey Starr, President of the Electric Power Research Institute, categorizes solar power as among our "speculative resources" with an implementation period of ten to one hundred years."[31] Not surprisingly, EPRI is an industry group that has heavily endorsed nuclear power, the radioactive option.

research, ERDA's funding decisions have been thoroughly imbued with the same favoritism the NSF has shown to large corporations. Many of these firms have held sizable military and aerospace contracts or made enormous commitments to nuclear energy. Some are involved in conventional fossil technology investments that could be adversely affected by solar development.

For example, as the prime contracts for its ten-thousand-kilowatt pilot facility for a large-scale solar thermal power plant, ERDA selected four colossi: Boeing Aerospace of Seattle, Honeywell Corporation of Minneapolis (makers of cluster anti-personnel bombs for Vietnam), Martin Marietta (a nuclear contractor) of Denver, and McDonnell Douglas of Huntington Beach, California. And among the sixteen organizations selected in 1975 by ERDA to develop low-cost solar arrays were Westinghouse Corporation, General Electric Corporation, Honeywell Corporation, Mobil-Tyco (subsidiary of Mobil Oil Corporation), Union Carbide Corporation (manager of federal uranium enrichment facilities) and Dow Corning (an offshoot of the infamous Dow Chemical Co., makers of napalm).*

The issue of providing federal support to corporations that might be tempted to drag their feet in the solar energy field was raised by James Piper, President of Piper Hydro Inc., in testimony to the Senate Select Committee on Small Business. Mr. Piper explained how conflict of interest could prevent a corporation with vested interests in the energy *status quo* from putting its best efforts into solar energy development:

> Westinghouse just submitted a proposal to the FEA for $1.75 billion for three atomic generating plants. If there was a dramatic increase of solar systems and a lowering of the need for electrical energy in the future, it is possible that they might not be able to sell those three plants for $1.75 billion. And that has got to be in the mind of the man in Westinghouse who decides whether to support solar systems or not . . . . There is not the profit, nowhere near it, in solar technology and producing equipment that there is in producing other types of energy equipment . . . .

Not only can the increased availability of low-cost solar energy reduce the overall growth in demand for conventional energy, but

---

*The complete list of organizations can be found in ERDA's *Weekly Announcement,* Vol. 1, No. 31 (October 29, 1975).

by making cheap alternative source available, it will place a ceiling on other energy prices charged by energy combines, making it harder for them to extort higher and higher prices from consumers.

By awarding big contracts to huge corporate war profiteers (with their long tradition of cost overruns for superweaponry) ERDA and the NSF deal setbacks to the innovative, independent and courageous small firms which pioneered the solar energy field and are now in danger of being deprived of their rewards. While these firms risked their slender capital resources and made technical breakthroughs with little or no government support, large grants to the behemoths of the corporate world now will assist the giants in muscling or buying out the little enterprises.

Regarding ERDA's solar funding efforts, Dr. Jerry Plunkett, president of Materials Consultants, a small solar firm, charged recently that the NSF funded expensive solar heat collector concepts and then used them "as examples as to why solar heating and cooling were so far away."[34] In one federal demonstration program option under consideration for fiscal years 1975-1979, ERDA was proposing to spend $109 million for the installation of solar systems in three hundred fifty residential and fifty commercial units.[35] The average cost per unit would be $272,500, which is far in excess of the amounts needed.

Other federal funding activities in the solar field have also impeded the development of cheap, universally accessible solar power for all. James Piper of Piper Hydro Inc. described how the Southern California Gas Co., which competed intensely against his small solar company, received a grant of $391,000 from the NSF that cost Piper Hydro potentially valuable business. NSF made the grant to Southern California Gas to install company-owned, solar assisted gas water heaters on apartment buildings and then to sell their heat to customers, ostensibly to test the feasibility of this idea. Piper told the committee he would have answered the $391,000 feasibility question *free* for NSF implying that there was no need for a feasibility study in that the basic feasibility question had already been answered long ago. Furthermore, the gas company said it would have to spend nearly $8,000 — half of it government money — for each water heating installation it completed. Piper has stated that his company has recently contracted to supply solar water heating, space heating, and electric cooling all for less than a fifth of that $8,000 cost per unit ($1,600). In the gas company project, Piper pointed out, "The tenants will not even own the water heaters, the gas company will. They will own the rights to the sun."

This type of grant boondoggle is not unique in the annals of federally funded solar research. The NASA Jet Propulsion Laboratory and Southern California Gas recently received NSF support to study ways in which utilities could provide money for the mass utilization of solar heating and cooling systems so that the utilities could then charge their customers for using the system.

If ERDA would use the growing appropriations for solar power so as to stimulate least-cost solar technology and insure it is made available directly to consumers, then in ten years or less we will be roofing or shingling our houses with solar electric cells. In most parts of the U.S., if just one square, twenty feet on a side on a south-facing rooftop, is covered with photovoltaic cells of just 15 percent efficiency, enough electricity could be produced for the average family. A square flat-plate collector area on the same roof thirty-two feet on a side could provide half to 70 percent of all the energy needed inside for space and water heating. This is the immediate challenge to which federal research dollars should be directed in the area of domestic solar utilization.

As a corollary to a saner disposition of federal solar funds, federal energy policy should be reoriented away from the bankrupt nuclear fission and breeder option towards a full-scale crash development program of solar power in every form. Some experts even believe that 100 percent of our electricity could be obtained from solar cells in just twenty-five years.[36] Surely such a promising and exciting potential deserves far more than the token federal support it is getting even now.

With modest encouragement from ERDA, the use of solar energy can become a well-diversified multi-billion dollar industry involving many firms, large and small, by 1985, without allowing electric utilities to exploit the sun at our expense. Conversion to a solar economy would also create a great many new jobs in manufacturing, installing, servicing, and repairing both domestic solar equipment and larger solar power plants. A wide range of opportunities would be created for skilled craftsmen and specialized contractors to ply their trades, while stimulating the industries that provide materials to those firms and individuals. Industries likely to benefit include manufacturers of glass, plastics, optical equipment, paint and selective coatings, pumps, pipes, valves, tanks, industrial chemical firms, and electronic control manufacturing firms. Jobs would be provided for architects, sheet metal workers, heating contractors, home remodelers, electricians, construction workers, plumbers, glaziers, and painters.

To encourage solar industries, ERDA ought to be promoting low technology solar power applications suitable for existing dwellings, instead of concentrating on developing nuclear power and high technology solar devices. One small step to assist in the implementation of cheap, decentralized solar power would be the preparation by ERDA of free manuals for the average homeowner, complete with construction diagrams explaining in lay terms how to build economical, efficient solar heating and cooling systems from cheap, readily available materials, and how to design new homes using simple, passive solar climate control.

If a fraction of the funds spent by the AEC extolling nuclear fission were devoted to films and printed media showing the step-by-step construction of efficient solar equipment, it would be of great assistance to the country in saving nonrenewable resources and liberating the economy from the burden of heavy oil imports.

*   *   *

Just as solar energy could gradually liberate the nation from dependence on nuclear power and foreign oil, it could also liberate average people from large corporate utilities and energy combines. For this as well as for environmental reasons, the universal solarization of the U.S. and the planet would be a most encouraging development. But ERDA and some utilities are delaying and obstructing it. Southern California Gas, as explained, has done its part to insure expensive solar energy by renting or leasing solar heaters and collectors to consumers. ERDA, in turn, has waylayed the development of cheap, decentralized and low-technology solar energy that consumers *could use now* in their own backyards and rooftops, without being ripped off by large energy conglomerates. Instead, ERDA has diverted its already meagre solar funding into high technology applications of solar energy, some of which are discussed in the next chapter. In short, ERDA and the big corporations are laying the groundwork so they can sell us the sun — at maximum profit for the energy combine. The most fervent and diligent citizen/community action efforts are going to be needed to oppose this ploy so that solar energy frees the average citizen from the energy monopolies.

SOURCES OF SOLAR INFORMATION

A number of useful plans for constructing your own solar equipment are listed in *Energy Primer, Solar, Water, Wind and Biofuels,* published by Portola Institute, 558 Santa Cruz Avenue, Menlo Park, Calif. 94025, and in Catalog 4A of the Environmental Action Reprint Service, 2239 East Colfax Avenue, Denver, Colo. 80206. Two other general sources that may prove useful are:

Carol Hupping Stoner (ed.), *Producing Your Own Power, How to Make Nature's Energy Work for You* (New York: Vintage Books, 1975), and

Stefan A. Szczelkun, *Survival Scrapbook #3 Energy* (New York: Schocken Books, 1974).

A few complete solar house plans are available, along with various types of solar equipment, from Edmund Scientific Co., 300 Edscorp Bldg., Barrington, N.J. 08007. Further information on the details of solar construction may be located by contacting the International Solar Energy Society (see address below). A more complete listing can be obtained from:

Solar Energy Industry Directory and Buyer's Guide 1975, published by Solar Energy Industries Association, Inc., 10001 Connecticut Avenue, N.W., Washington, D.C. 20036. 202 293-1000. The National Intervenors, 153 E Street, S.E., Washington, D.C. 20003, also has a list available.

Incidentally, the Solar Energy Industries Association gave its "Man of the Year Award" in 1975 to Congressman Mike McCormack, perhaps the most ardently pro-nuclear person in the U.S. Congress. McCormack once worked at the AEC's Hanford facilities and is a vociferous proponent of nuclear power and breeder reactors. He sponsored the National Solar Heating and Cooling Demonstration Act of 1974 which some solar proponents claim was intended to hinder solar power's development.

People interested in further study of solar energy may consult a short solar energy bibliography prepared by the International Solar Energy Society, American Section, 12441 Parklawn Drive, Rockville, Md. 20852. The bibliography is also available at pp. 1521-2524 of the Senate Select Committee hearing record cited above and in the notes to this chapter. (The 5-volume hearings can be bought from the U.S. Government Printing Office, Washington, D.C. 20402.) The Select Committee hearing record also contains an annotated solar energy bibliography (pp. 3727-3751, dated Sept. 1974). Another bibliography on solar and other energy technologies is available from Alternative Sources of Energy, Route 1, Box 36B, Minong, Wisc. 54859. Finally, an excellent and classic authoritative reference on the fundamentals of solar energy is Farrington Daniels, *Direct Use of the Sun's Energy* (New York: Ballantine Books, 1964).

NOTES

1. Shurcliff, W. A., "Solar Heated Buildings, A Brief Survey," Cambridge, Mass. Available from *Solar Energy Digest*, P.O.B. 17776, San Diego, Calif., 92117, $7.

2. The President's Materials Policy Commission.,Final Report, *Resources for Freedom, Vol. IV, The Promise of Technology* June 1952, p. 220.

3. Seamens, Dr. Robert C. Jr., Speech prepared for delivery to Atomic Industrial Forum, St. Francis Hotel, San Francisco, Calif., November 18, 1975.

4. Steadman, Philip, *Energy, Environment and Building* (Cambridge, England: Cambridge University Press, 1975).

5. Seifert, W., et. al., eds., *Energy and Development — A Case Study*, MIT Report No. 25 (Cambridge, Mass.: MIT Press, 1973).

6. Inter-Agency Technical Review Panel, Sub-Panel IX, *Solar Energy Program Report*, December 1973.

7. *Solar Energy Research Program Alternatives*, Mitre Corporation, McLean, VA, December 1973.

8. Halacy, D. S., *The Coming Age of Solar Energy* (New York, N.Y.: Harper and Row, 1963). Avon paperback, 1973.

9. Clark, Wilson, *Energy for Survival* (Garden City, N.Y.: Doubleday & Co., 1975).

10. Hill, Gladwin, "Rapid Gains Seen for Solar Energy," *New York Times*, Aug. 3, 1975.

11. Select Committee on Small Business, U.S. Senate, 94th Congress, *Hearings, Energy Research and Development and Small Business*, pp. 37, 56, 64, GPO, Washington, D.C. 20402, ($6.60).

12. Bentley, Alene E., "'Harnessing' the Sun," *National Enterprise*, Vol. 5, No. 9 (September 9, 1975).

13. Steadman, *op. cit.*

14. Mlavsky, A. I., "The Silicon Ribbon Solar Cell — A Way to Harness Solar Energy;" Tyco Laboratories, Inc., Waltham, Mass. 02154, June, 1974.

15. *Solar Energy Program*, ERDA-15, Division of Solar Energy, U.S. Energy Research and Development Administration, March 1975 (pamphlet). See also Select Committee, *op. cit.*, pp. 174-191.

16. *Solar Energy as a National Resource*, National Science Foundation/National Aeronautics and Space Administration Solar Energy Panel, December, 1972.

17. *Inter-Agency Technical Review Panel*, *op. cit.*

18. *Solar Energy Research Program Alternative*, *op. cit.*

19. *Solar Photovoltaic Energy Hearings*, Subcommittee on Energy, Committee on Science and Astronautics, U.S. House of Representatives, 93 Congress, 2nd Session, June 6 and 11, 1974 (No. 43) USGPO, Washington, D.C.

20. *The Role of Small Business in Solar Energy Research, Development, and Demonstration*, Interim Report, Select Committee on Small Business, U.S. Senate, October 7, 1975, USGPO, Washington, D.C., 1975.

21. Clark, *op. cit.;* see also *Hearings, op. cit.*, pp. 3412-3413.

22. Daniels, Farrington, *Direct Use of the Sun's Energy* (New Haven, Conn.: Yale University Press, 1964); paperback ed. available from Ballantine Books, N.Y.

23. Select Committee, *op. cit.*, p. 2375.

24. "Bright Ideas for Solar Cell," *San Francisco Examiner*, June 26, 1975.

25. Smith, Otto J.M., "Smith Multimodule Solar Electric Plant," Monograph, 1975. 612 Euclid Avenue, Berkeley, Calif. 94708.

26. Smith, *op. cit.*

27. von Hippel, Frank, and Robert H. Williams, "Solar Technologies," *Bulletin of the Atomic Scientists*, November, 1975.

28. Hottel, H. C. and J. B. Howard, *New Energy Technology, Some Facts and Assessments* (Cambridge, Mass.: The MIT Press, 1971).

29. Bryant, Alden "Statement to the Senate Committee on Public Utilities, Transit and Energy, California Legislature, March 23, 1976. (Bryant is an engineer and economist with the Northern California Solar Energy Association.)

30. Select Committee, *op. cit.*

31. Starr, Chauncey, "Energy and Power," *Energy and Power* (San Francisco, Calif.: W. H. Freeman & Co., 1971).

32. *Solar Energy Projects of the Federal Government*, FEA/C-75/247, Federal Energy Administration, National Energy Information Center, Washington, D.C. 20461, January, 1975.

33. "The Question of Large Company Domination of the Solar Energy Industry," Office of Statistics, Federal Energy Administration, 1975 (paper).

34. Select Committee, *op. cit.*, p. 71.

35. "Statement of Dr. John M. Teem," Assistant Administrator for Solar, Geothermal, and Advanced Energy Systems, ERDA, before Senate Select Committee, *Hearings*, pp. 193-223.

36. Morris, David, "The Dawning of Solar Cells," Institute for Local Self-Reliance, 1717 K St., N.W., Washington, D.C. 20009.

12

OTHER ENERGY OPTIONS

This final chapter on alternatives to nuclear fission highlights several alternative power options, some available today, others still under development. We will consider geothermal power, wind power, ocean thermal power, tidal and wave power, fusion power, biomass power, oil shale, and cleaner coal technologies. Because such a broad subject area is being covered, the treatment of each technology will be limited to a brief description of the technology, an explanation of its uses, the amount of energy possible from it, the probable costs of extracting the energy, and any special relevant technological, environmental, or economic problems.

Greater reliance on crude oil and natural gas as an alternative to nuclear fission is not considered here because our dwindling domestic oil and gas resources require that we get a smaller proportion of our total energy from these fuels during the next decades than we have in the past, unless the unacceptable economic burden of vastlly expanded fuel imports is imposed on the nation. More attention in this chapter is devoted to non-fossil-fuel technologies than to oil shales or to modern coal conversion technologies, in which coal is processed into a liquid or gaseous fuel. Those two fossil-fuel technologies are amply described elsewhere in the energy literature, and both have powerful advocates among energy suppliers with large economic stakes in them. High monetary costs, severe net energy penalties, and devastating environmental effects render oil shales and some coal conversion proposals unappealing. Consequently, they are not recommended here as alternatives to nuclear power.

GEOTHERMAL POWER

Scores of fleecy plumes arc skyward only to be seized and devoured by green demons that haunt the boughs of imperial conifers; bundles of silvery bullwhips, cracked by an invisible giant who lurks behind the western hill, are caught in stop-action as they rise and fall in unison . . . . It is an odd amalgam of technology and nature, of the Tin Woodsman of Oz and the Sorcerer's Apprentice, gently underscored by the whispering, slightly syncopated whuff-whuff . . . whuff. . . whuff of the wellhead silencers.

— Robert C. Axtman, "Environmental
Impact of a Geothermal Power Plant,"
Science, Vol. 187, No. 4179, March 7, 1975.

As tourists, many of us have marveled at the geysers of Yellowstone National Park in Wyoming or the hot bubbling mud pots of Mount Lassen National Park in California, without realizing that similar sources of underground heat can be used to generate electricity.

Geothermal energy is heat derived from the magma, the molten rock deep within the earth. In regions of intense seismic activity where the earth's crust is fractured, magma wells up toward the surface. When magma or magma-heated rock comes in contact with ground water that has seeped down through porous regions of the earth's crust, the heat turns the water to steam. Some of the steam collects in nonporous, impermeable regions deep underground where it is capped by geological formations. It may remain there under pressure or it may seep to the earth's surface with hot water, as in a geyser, or mixed with hot gases, as in a fumarole. If wells are drilled into underground hydrothermal reservoirs, large amounts of the steam can be brought to the surface. Geologists have found that geothermal sources of hot water are much more common than underground sources of dry steam. Hot dry rock is even more prevalent, existing everywhere at varying depths beneath the earth's surface. One need only be willing to drill deeply enough.

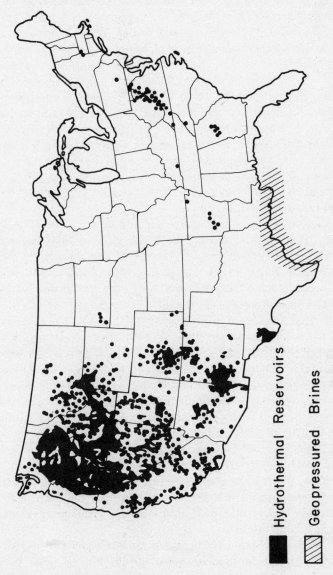

Figure 12-1. Distribution of U.S. Geothermal Resources

Although officials of large energy companies shrilly warn of an energy crisis, their researchers have scarcely reckoned our vast geothermal resources, and the federal government has supplied only minimal funds to develop geothermal power. Yet U.S. geothermal resources are enormous and their exploitation depends only on extending existing technology to solve mundane problems like corrosion of machinery from hot geothermal brines and liquid waste disposal. These impediments can be solved with low-technology and do not depend on any major new scientific breakthroughs. The most serious technical and economic problems liable to impede geothermal development are the costs of deep wells and of wear-resistant generating machinery, and the significant environmental impacts that may accompany geothermal development. These include possible air, water, and noise pollution, as well as possible land subsidence and induced seismic disturbances.

Geothermal energy is already used for electric generation in the U.S., Italy, Japan, Mexico, the U.S.S.R., and New Zealand, where in the latter case about 7 percent of the country's power comes from a single geothermal plant.[1] New Zealand now gets about the same percentage of its electricity from geothermal sources as the U.S. gets from nuclear fission.

The first significant use of geothermal power for electrical generation was in Larderello, Italy, in 1904. This geothermal field soon became the primary source of power for Italy's railroads. Geothermal heat has been used in Iceland since 1943. The city of Reykjavik is almost totally heated by pipes leading from a geothermal hot-water source, eliminating the need for furnaces and space heaters. Reykjavik's geothermal energy supply is so plentiful that it is used to heat swimming pools and greenhouses where grapes, tomatoes, and bananas are grown and to power food and mineral processing industry. Of course, not many cities have such accessible geothermal sources. However, as geothermal technology improves, more geothermal sources that are now too deep in the earth's crust, or are otherwise unsuitable for economic exploitation, will become usable.

If geothermal steam is hot, clean, and relatively dry (vapor-dominated), it can be used to spin a turbine directly, after rock particles and dust are removed in centrifugal separators. Other geothermal sources produce steam and hot water together, but the

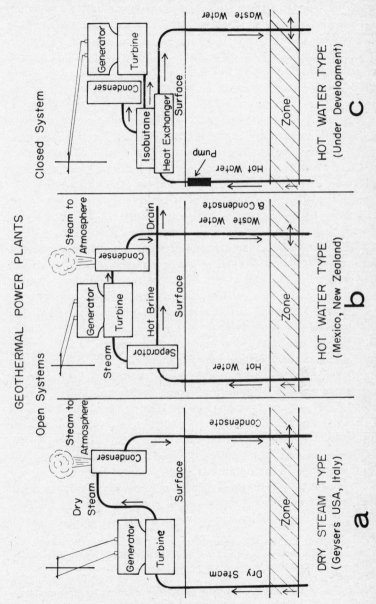

Figure 12-2. Geothermal Power Plant Types

water is so laden with dissolved mineral impurities that it would rapidly corrode and wear out a turbine. Researchers believe that this problem can be solved by sending the steam to a heat exchanger. There, its heat can turn clean water or a secondary fluid (such as isobutane) into vapor to drive the turbine. The hot, briny fluid need never come into direct contact with the clean water or the turbine.

Although all geothermal plants produce their power from steam today, studies currently are being conducted in the U.S. and elsewhere to overcome the barriers to using hot brines and hot rocks.[2] That breakthrough would greatly increase the power available from geothermal sources and would considerably expand the areas suitable for geothermal development. Geothermal power at a given source may be increased as much as 60 percent by withdrawing energy from the hot brines found with steam.[3]

Economically competitive power from hot dry rock one day also may be obtained. Water would be pumped into a well drilled down to the rock, and steam would be recovered from another well sunk nearby; the steam would power a turbine.

Batelle Memorial Institute has been exploring the potential of hot dry rock as a geothermal source in Marysville, Montana. Noting that this resource may exist throughout the western U.S., a Batelle official indicated that the energy potentially available from hot rock alone might equal the energy in the entire U.S. coal reserve.[4]

Large amounts of geothermal power one day also may be available for the southern U.S. Dr. Eugene Herrin of Southern Methodist University has concluded that an immense reservoir of hot water mixed with natural gas exists under the Texas flatlands of the lower Gulf Coast. He believes the energy there may exceed the total energy of all the oil and gas ever obtained from the region.*

The great advantage of geothermal power is that no fossil fuel is needed in the operation of the plant. Because no boiler is used (as in a conventional fossil-fuel plant), the capital costs of

---

*The extraction of geothermal energy from hot brines is currently being studied under an ERDA contract by researchers from the Lawrence Livermore Laboratory using a well operated by Southern California Edison, Southern Pacific Land Co., and Phillips Petroleum Co., at El Centro, California. ERDA and the San Diego Gas & Electric Company are currently engaged in building an experimental ten megawatt geothermal pilot plant to use hot brines at Niland, twelve miles from El Centro, California.

the geothermal plant may be only half, or less, of a fossil-fired unit.[5] The main costs are for plant construction and drilling.

The only geothermal electrical plant in the U.S. today, and the largest in the world, is located at The Geysers in Sonoma County, California. Operated by the Pacific Gas and Electric Co., eleven geothermal wells generate about 500 MW of electricity — enough to meet half the electrical needs of San Francisco. The ultimate potential of this field alone has been estimated at ten times its present output.[6] PG&E already has plans to expand the field's production to 908 MW by 1977.

In the U.S., at least 20,000 MW of electrical power could be generated from geothermal wells by 1985,[7] an output equal to 5 percent of current U.S. electrical capacity. However, the size and value of U.S. geothermal reserves is not yet accurately known, and exploration for geothermal sites has so far been minimal. Nonetheless, hot water resources have been found in California, Idaho, Nevada, New Mexico, Oregon, Texas, and many other states. Such "wet" geothermal source power may one day be used to produce desalinated water for irrigation as well as power.

Large-scale electrical production is not the only use for geothermal energy: for years some houses in Klamath Falls, Oregon and many offices in Boisie, Idaho, have been heated with hot ground water and steam.

Geothermal energy is not drawn from the earth today without environmental penalties, depending on the technological refinement with which the extraction is done, and on the chemical composition of the fluid withdrawn.

For example, the Geyser geothermal field is set in a rugged canyon filled with active seeps of hot water and steam that bring brilliantly colored clays to the surface. Prior to large-scale drilling at The Geysers by a consortium of Union Oil Co., Magma Power Co., and Thermal Power Co., the area had an eerie, primeval, unspoiled appearance, as warm natural vapors wafted across the gravel-strewn terrain. Now the landscape has been marred by drilling rigs, bulldozers, construction crews, long-distance power lines, and the plant itself. Hillsides have been scarred by roads and leveled by drilling pads; streams nearby have been contaminated with drilling muds, detergents, and petroleum products. The canyon's pristine solitude and the tranquil beauty of Big Sulphur Creek, which flows past the plant, have been lost in the quest for power. The Big Sulphur Creek trout fishery, once a valuable steelhead spawning tributary of the Russian River, has been largely ruined. (In 1974, the State Department of Fish and Game counted less than a hundred juvenile steelhead in the creek, instead of the usual twenty thousand.*

Traces of naturally occurring radon-222 gas have also been detected by PG&E in a study of the discharges from The Geysers' wells. The company says the quantities are negligible, but at this writing, that remains to be confirmed by independent analysts.[8] The long-term environmental effects of radon emissions from large-scale geothermal developments are not currently well known.

The environmental effects of a geothermal installation have been carefully documented in studies of the Wairakei geothermal plant near Lake Taupo in New Zealand. In contrast to the Geysers plant, the hot waste water is not reinjected into ground wells. Instead, water containing large amounts of dissolved carbon dioxide and hydrogen sulfide is discharged directly into the Waikato River near the plant.[9]

Hydrogen sulfide, the gas that gives rotten eggs their odor, is detectable even in extremely small concentrations (.002 ppm) and in larger doses can cause eye and throat irritation. In very large concentrations (600 ppm) it can be fatal. The plant's hydrogen sulfide releases are sufficient to give the river a strong odor and to kill rainbow-trout eggs. Twenty-six chemical effluents have been identified at Wairakei, where visitors are told the plant has no environmental effects. The effluents include arsenic, mercury, and silica. Even without biological concentration — which can intensify poisons many thousands of times — arsenic from the plant at times accumulates in the river at concentrations exceeding those permitted in U.S. drinking water. Trout from the river have already been found to have high levels of mercury in their flesh.

In a few respects, a geothermal plant may be more environmentally harmful than a fossil-fuel plant. Discharges from a particular geothermal field depend on the geochemical characteristics of the hydrothermal reservoir from which the power comes. Conse-

---

*To avoid even more serious environmental consequences at The Geysers, PG&E reinjects the hot waste water from the plant back into ground wells. The company is also building emission scrubbers to remove toxic hydrogen sulfide gas from the vapors in the plant's cooling towers.

quently, carbon dioxide from a geothermal plant can be as much as ten times greater than from an ordinary thermal plant. And because geothermal plants operate at relatively low thermal efficiencies compared to fossil plants, the geothermal ones produce somewhat more waste heat per unit of power generated. But some of the environmental effects of geothermal plants can be greatly mitigated. As noted previously, the reinjection of hot waste water and gases back into the earth shields the atmosphere and local water drainage systems from plant effluents and may lessen the chance of ground subsidence. It can also replenish the geothermal reservoir itself and postpone its depletion.

ECONOMICS

For a long time PG&E produced power from the Geysers' steam at a low cost — 5 mills per kWh — less than from the company's Humboldt Bay Nuclear Plant. Then Union Oil joined the Magma Power-Thermal Power consortium operating the Geysers facility* and PG&E agreed to a new precedent-setting contract for geothermal steam, coupling its cost to the escalating cost of oil. The escalator clause is undoubtedly a great advantage to the oil companies, which are investing in geothermal resources, because it protects them from underselling their own oil. The effect of the oil companies' extensive geothermal holdings coupled with Union's precedent-setting pricing policy is to deprive consumers of a potentially cheap source of energy. In California's Imperial Valley — an area likely to be of great geothermal importance in the future — 70 percent of the land with geothermal potential is held by four companies: Standard Oil, Union Oil, Southern Pacific Land, and Magma Power.[10] As many as twenty-five thousand acres of geothermal land can be leased perpetually from the government for as little as a dollar per acre. This gives the leaseholder — in California, usually a major oil company — exclusive legal rights to develop the geothermal resource, or to withhold it from development.

Large U.S. energy suppliers are highly cognizant of the opportunities to minimize interfuel competition, now and in the future.

---

*Natomas, the multinational conglomerate that struck it rich with Indonesian oil, bought Thermal Power Company in 1974, thereby acquiring 25 percent of the Geysers.

In a *Los Angeles Times* feature story, George Alexander told how, "with little fanfare, but with great purposefulness and skill," major U.S. oil firms are dominating not only the geothermal resources of California, but of the entire U.S.[11]

Speaking at the second world conference on geothermal energy in San Francisco during May 1975, California Lieutenant Governor Mervyn M.Dymally charged that the oil companies are intentionally developing their geothermal leases more slowly than necessary because they can make higher profits by selling oil rather than geothermal power.[12] Oil company delegates to the conference countered by citing environmental and technical difficulties in getting geothermal fields into production, and claimed that government regulations also were responsible for the slow pace of development.

If the environmental problems of geothermal power are carefully handled and a major geothermal development program is launched by ERDA, geothermal energy could well prove to be a major source of power for the nation. Research in this field ought to be vigorously funded, and the lease-hold system that permits large firms to stalemate geothermal development ought to be overhauled.

WIND

Windmills have done useful work for at least a millennium, since the early artisans of tenth-century Persia built machines to convert the momentum of the winds into mechanical work. Some of these forerunners of today's windmills pumped water; others used horizontally turning sails mounted on a vertical axis atop a square tower to turn millstones for grinding corn.[13] From Persia, use of the windmill spread throughout the world, primarily for milling grain and pumping water. Use of windmills to operate mechanical pumps was common even in the U.S. until quite recently, and windmills for this purpose are still produced on a small scale. About 175,000 water pumping mills currently exist in the U.S., half of them in working or repairable condition.[14] But the wind machines of primary interest from an energy-resources standpoint are wind generators (also called turbines) used to produce electricity.

Widespread commercial use of wind generators in the U.S. began in the 1920s. These machines used blades that rotated on a horizontal shaft and transmitted power through a gearbox to an electric generator that produced direct current. The gears enabled the generator to spin at higher rates than the shaft.

Fifty thousand small wind turbines were producing electricity in the Midwest as recently as the 1950s, at which time the impact of the Rural Electrification Administration — launched by Franklin D. Roosevelt in the 1930s — finally made them uneconomic.[15] Since the steep price rises in electricity of 1972-1975, American interest in wind turbines has intensified.

Wind power today is technologically feasible in systems suitable for small-scale up to multimegawatt-scale power generation, but high cost and the necessity of access to an appropriate tower site (not a rooftop), prevent its general use for single-family dwellings. The first cost of a small-scale commercially available wind generator plant, its storage batteries, installation, and AC converters (or purchase of DC-operated appliances) is not economically competitive with commercially available power from utilities for most people except in remote areas of the U.S. Prices for small-scale units begin at about three thousand dollars, without installation, and go up from there. The cheaper units often require frugality in power consumption, and all units are suitable only in areas with adequate wind speeds close enough to ground level to make tower costs economic. Rooftops are unsuitable sites for wind generators in most cases because of the unacceptable stresses on the roof, and the likelihood that noise and vibration will disturb the residents.

Homemade wind generating systems can be constructed and several of the references cited at the end of chapter 11 offer construction plans. However, reductions in costs achieved by home construction tend to be gained at the expense of durability under the high stresses to which wind systems are exposed, or at the expense of high-efficiency performance, or both. But to those for whom a high first cost is no object, wind power can be the total source of domestic electrical supply or a valuable supplement to it. Equipped with a twenty-five-foot diameter rotor, a wind generator can produce enough power in many parts of the U.S. to meet the power needs of an all-electric home. A ten-foot diameter wind generator can produce enough electricity every day to operate a pollution-free electric car suitable for urban transport.[16] (Such vehicles are feasible but not yet commercially available to most consumers.)

Commercial wind power systems tend to be more economic for communities in which the capital cost of the wind generating plant can be shared by its beneficiaries. And the economic competitativeness of such systems depends on factors relevant to the particular situation, including site characteristics and the costs of alternative power supplies. But large-scale wind systems have a potential widely recognized in the scientific community for generating electricity at economically competitive prices.

As any student of wind power knows, there is no energy supply crisis in the U.S. The wind energy available above the forty-eight contiguous states is many time the current U.S. energy use. Within twenty-five years, according to an NSF/NASA estimate, almost as much electricity could be generated from the wind as the U.S. consumed in 1972.[17]

Areas suitable for large-scale wind power development in the U.S. include large areas of the Great Plains, the eastern foothills of the Rocky Mountains, mountainous regions of New England,

NASA photo

Figure 12-3. ERDA's 100 kW windmill on a 125 ft. tower near Sandusky, Ohio.

the Aleutian arc, the Texas Gulf Coast seaboard, and the continental shelf off the Northeastern seaboard.[18] Winds are generally much stronger several hundred feet above the ground than at ground level, and velocities in the range of twenty miles per hour can be found at such heights over large areas of the U.S. Winds at sea are on the average even more powerful.[19]

The power available from the wind depends on the air density, the area intercepted by the wind turbine's rotors, and the wind velocity. The power varies with the square of the wind generator blade's arc or diameter, and with the *cube* of the wind velocity. This implies a large power dividend can be reaped by increasing the blade diameter, and that a much larger dividend can be had by utilizing stronger winds. As rotor diameter doubles or triples, for example, power increases four and nine times, respectively. As wind velocity doubles or triples, power increases eight or twenty-seven times, respectively. In addition to these power-determining variables, wind turbines obey other laws of physics that limit the maximum theoretical efficiency a wind generator can achieve, in converting wind to mechanical energy, to 59 percent. However, because of aerodynamic blade inefficiencies and other mechanical energy losses, the actual efficiency achievable is closer to 35 percent.[20] Another physical fact limiting the output of windmills is the wind's intermittency. So typical wind generators on the average produce power at only 15 to 25 percent of their rated capacity. Therefore, to provide an average amount of power equivalent to that of a nuclear plant operating at 60 percent capacity, three times as much wind generating power capacity has to be installed. This naturally increases the overall cost of wind-generated electricity. If an energy storage system is provided, this will also raise costs. (Storage systems will be discussed later in this section.)

Although wind is sometimes thought of as a diffuse source of power, the indirect solar energy available from the wind via the expansion and contraction of our atmosphere is recoverable in a more concentrated form than direct solar energy used to produce electricity. The wind power reaching a wind generator perpendicular to a twenty-mile-per-hour wind is 45 watts per square foot.[21] If one third of the energy can be converted to useful work or electricity, then 15 watts are available per square foot. By comparison, the solar flux on the ground each day is 64 watts per square foot; if converted to work by a solar collector with a 10 percent efficiency, only 6.4 watts will be available. Consequently, the wind system can produce far more power per square foot than the solar collector where there are strong prevailing winds.

Wind power systems have many advantages over fossil-fuel technology or uranium fission power. Wind systems consume no fuel so they produce no poisonous or foul-smelling effluents, no flyash, no thermal pollution, no routine radioactive emissions, and no radioactive waste. Nor do they present any risk of catastrophic core meltdowns. These are strong incentives to use wind power.

The main technical obstacles to the large-scale generation of wind electricity are that tens of thousands of very large machines, hundreds of feet tall, might be deployed, and their intrusive metallic hulks would mar the landscape (unlike the picturesque water pumping machines of the Netherlands). Yet the know-how to build tall wind turbines is available, and they could be put out at sea or in locations where few people would be disturbed by them.

The technology for producing commercial electric power from the wind has been available for at least thirty years. A 1.25 MWe wind generator was built on a hilltop known as "Grandpa's Knob" in Rutland, Vermont, and it supplied power to parts of Central Vermont from 1941 to 1943. Designed by engineer Palmer C. Putnam, this machine had 175-foot diameter stainless steel blades and was built as a commercial prototype to assess the economic feasibility of larger machines. However, a broken blade ended the useful life of the machine in 1945 (it was immobilized two years earlier by a burned-out bearing). A crack in one blade had developed because of faulty assembly techniques, use of a new alloy, and the totally avoidable bearing failure. When the blade broke, it was already scheduled for replacement and would have been replaced earlier except for the scarcity of metals due to the war.[22]

Despite the untimely end of the Grandpa's Knob machine, its operation as a first prototype, large-scale wind generator was a success. The effort showed that the major technical problems associated with the generation of large amounts of power from the wind could be solved, and the experiment itself came within a factor of two of being economic. Moreover, the machine took only two years to build, from the time its builder first became interested in designing the plant. Nonetheless, because windpower in Vermont then had what seemed like economically unbeatable competition from cheap water power and cheap power from out-of-state utilities, S. Morgan Smith Co. of Pennsylvania, which built the machine, decided not to go into commercial production.

A few years after the Smith-Putnam machine, a 6.5 MW windmill was commissioned by the Federal Power Commission and was designed by engineer Percy Thomas. A bill to provide federal funds for construction of the Thomas machine died in a congressional committee in 1951, however, at a time when Congress was feeling pressured by the many financial obligations of the Korean War.

In view of the results obtained by wind energy pioneers in the 1940s and as a result of recent advances in solid state electronics, aerodynamics, energy storage technologies, materials sciences, mass-production techniques, and computerized design-optimization techniques, it seems plausible that the economic and technical problems encountered by earlier wind plant developers today can be conquered. The primary remaining objections to the large-scale deployment of wind power are the size of wind generators necessary to produce commercial power, the number of machines that would have to be deployed, and the "visual pollution" they might create.

To equal the generating capacity of a thousand-megawatt nuclear plant, roughly eight hundred windmills with two-hundred-twenty-five-foot diameter blades would be required.[23] To produce all the electricity currently used in the U.S., some one million windmills with two-hundred-foot diameter blades would be needed.[24] If stood end-to-end, these windmills would stretch for forty thousand miles.

For various reasons, these facts do not bother wind energy enthusiasts. Professor William Heronemus, a civil engineer from the University of Massachusetts at Amherst, has been a dedicated advocate of large-scale wind electric systems for several years. Professor Heronemus believes that an offshore wind power system (OWPS) can be built off the East Coast to provide all the electricity needed for the six New England states more cheaply than by nuclear power.[25]

Heronemus estimates that the total cost of installing such a wind system, complete with hydrogen energy storage, would be about $22.4 billion. Hydrogen can readily be produced from seawater by electrolysis, the passage of an electric current through the water, causing the dissociation of water into oxygen and hydrogen. The product gases can be collected at opposite terminals of an apparatus. Direct current generated by the wind would be used to perform the electrolysis, and the resulting hydrogen and oxygen would be pumped ashore where they could be recombined as needed in a fuel cell to produce electricity.

Delivery of the wind system's components could begin within twenty-four months of the program's inception, Heronemus states, and several large corporations are capable of producing the hardware. Wind generator construction does not require advanced technology (unlike fusion and breeder reactors), and the wind generator is readily adapted to mass production techniques. Professor David R. Inglis, of the University of Massachusetts, another supporter of massive reliance on wind power, believes that six thousand six-megawatt wind machines could be built offshore from New England. Professor Inglis agrees with Professor Heronemus that wind power would be practical in conjunction with hydrogen as an energy-storage system.

Solar and wind power systems can also be effectively integrated. The energy potential available from combined direct solar and wind power systems was studied by Dr. Bent Sørensen, a physicist at the Niels Bohr Institute in Denmark. A great deal of practical experience with wind generators has been accumulated in that country since Professor Poul La Cour used wind energy to electrify the Danish village of Askov around the turn of the century. Professor Sørensen recently concluded that all Denmark's energy needs could be met by solar and wind power by the year 2050, and that over twenty-five years the system's costs would be comparable to those of other energy systems. He views the project's construction as a valuable way to help solve Denmark's 10 percent unemployment and he believes that wind power on Denmark's west coast is economically competitive with nuclear power, and is "probably substantially cheaper." [26]

Professors Inglis and Heronemus are no longer alone in their beliefs that wind power can be economic soon. According to a baseline study by W. L. Hughes and others, once mass production of wind generator units has begun, wind power systems will probably cost one hundred fifty to two hundred dollars per kilowatt in 1974 equivalent dollars. [27] Plants on the lower end of this cost spectrum would clearly be competitive to nuclear plants with capital construction costs of five hundred dollars per kilowatt, even if the installed capacity of the wind units had to be three times the installed capacity of the nuclear plants, to make up for the winds' variability.

An even lower estimate of wind-energy-system cost has been made by R. Ramakumar and colleagues who estimated that wind energy systems could be built for one hundred twenty-five to one hundred fifty dollars per installed kilowatt. [28] They believe such plants would be competitive with coal power and very competitive with nuclear power. This estimate is consistent with the conclusion made by Heronemus in 1972 that fifty-foot diameter wind generators could be built for one hundred dollars per kilowatt, assuming a production of twenty thousand units a year. [29]

Although the wind power systems just discussed include provisions for energy storage either through hydrogen production, compressed air, or other methods, wind energy can be useful for baseload power even without energy storage systems, as long as the units are dispersed over large enough geographic areas so that when the wind is not blowing at one site, other turbines in the system are delivering power. In such a baseload system, energy to meet peak demands could be supplied by existing fossil-fuel plants.

Barring objections to the physical appearance of large wind generators, wind energy "farms" could be situated in the Midwest where as much as forty megawatts could be produced per square mile of land. [30] Thus a square plot eight and two thirds miles on a side could contain three thousand megawatts of installed capacity. If the soil were arable and water were available, high-yield farming and cattle grazing would be compatible land uses. A limited number of these sites could probably be found in areas of low population density and low scenic value, but a better solution would be to place large wind turbines at sea.

Although the technology to build multimegawatt wind generators has been available for a quarter of a century already, the AEC and now the Energy Research and Development Administration have made only faltering gestures, so far, to expeditiously develop large-scale wind power. Perhaps ERDA, with its inherited profission bias, will try to delay viable alternative energies like wind power long enough for uranium fission technology to become even more entrenched than it already is. ERDA and NASA recently dedicated a hundred-kilowatt windmill in Sandusky, Ohio, primarily "for testing and experimental purposes."[31] In this project, the U.S. energy establishment took a flying leap thirty five years into the past.* This "exploratory" wind generator is less than a twelfth the size of the Smith-Putnam machine in Vermont. In addition to commissioning the Sandusky project ERDA should have simultaneously commissioned the construction of several types of megawatt-scale wind generators and prototype energy-storage systems. At the present pace of ERDA's wind energy program, it may be a decade before a wind-energy system of even a hundred megawatts is constructed! This is the same intentional procrastination ERDA is exhibiting in its reluctant development of most nonnuclear energies. The effect is to sabotage the development of safe energies for the near future.

OCEAN THERMAL

Few nonscientist have ever heard of Ocean Thermal Power (OTP), yet this indirect form of solar energy may prove to be the energy "sleeper" of the year. Its theoretical potential as an energy source is impressive, although OTP plant performance needs to be verified by experiments with prototype plants.

The ocean is a gigantic reservoir of solar energy. It comprises 71 percent of the globe's surface and receives a total of almost two and a half times as much solar radiation as do the land masses. Yet the energy of this phenomenal resource is not put to direct use for power generation at all, except in one tidal energy plant at La Rance, France. No OTP plants are currently in operation, and the largest of the experimental plants ever built did not prove economic when tried in 1947 off what was then the Gold Coast of Africa.

---

*In its announcement of the dedication, ERDA claimed, "The Sandusky wind turbine is the largest wind energy system now in operation and is the second largest ever built." Although acknowledging the Smith-Putnam effort, ERDA failed to note that a hundred-kilowatt machine was built by the Soviets in 1942, and that two-hundred-kilowatt machines were built by the Danes in 1942 and in 1957. Other hundred-kilowatt machines were built in Germany and the United Kingdom.

OTP plants utilize the large temperature differences that exist in the waters of tropical oceans to power a heat engine. Water on the ocean's surface between the Tropics of Cancer and Capricorn is about 77°F, while far below the waves of the same seas flow dark cold currents of 40° F. The technology of OTP plants is basically simple. The plant has separate intake pipes for warm surface water and for deep polar currents. Inside the plant, the warm sea water or a volatile fluid (such as propane or Freon), with a lower boiling point than water, is turned to vapor. Cold water is pumped up from the ocean depths through huge pipes and is used to help draw the expanding vapor through a turbine. The turbine turns a conventional generator and makes electricity. The cold water then condenses the warm vapor so it can be recycled to the hot end of

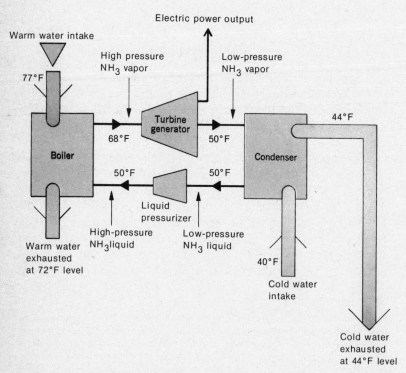

Figure 12-4. Ocean Thermal Power Plant.

the plant for warming and reexpansion. The engines' thermodynamic efficiency is quite low (about 6 percent) but since it uses no fuel once in operation, this is not a fatal defect.

Although most potential OTP sites are in the tropics, the Gulf Stream along the Eastern Coast of the U.S. from Florida to North Carolina is suitable for OTP development, and it could supply many times the energy needs of the U.S. as projected for 1980.

Various kinds of OTP plants have been proposed; although some designers envision OTP plants on shore, most OTP engineers propose to produce OTP with offshore plants. These structures would be submerged, parts of them at depths of 150 feet or more below the surface, to avoid the stresses of wind-water turbulence. Even in hurricanes, the ocean stays relatively calm at such depths.

Almost none of the drawbacks to land-based solar power apply to OTP. Because the ocean surface is a colossal, ready-made solar energy collector whose temperature is relatively invariant, conventional solar collectors do not have to be built to gather solar heat. Also, because the ocean waters store the sun's heat from season to season and from day to night, no heat storage apparatus is necessary. The plants can produce power twenty-four hours a day, regardless of transient atmospheric conditions affecting the transmission of sunlight. Only slight losses in efficiency would occur during cloudy weather. Furthermore, OTP installed capacity need not be a multiple of rated power capacity as with wind generating systems. A great advantage of the OTP system is that its fuel costs are essentially zero, as with other solar technologies, although relatively small amounts of fuel would be necessary for pump priming during start-up.

The power potentially obtainable from networks of OTP plants is tremendous. The *Solar Energy Program Report*, prepared for the AEC by the NSF,[32] states that there is enough warm surface water in the world's tropical oceans to produce as much as a thousand times the current electrical generating demand of the U.S. The panel concluded in 1973 that the cost of building OTP plants pursuant to the development of the requisite technology would be about four hundred dollars per kilowatt (in 1973 dollars). This estimate is predicated on the assumption that the federal government would support research and development of OTP through the creation of successful demonstration models. Other estimates of OTP cost are even more optomistic. Engineers James H. Anderson, Jr. and Sr. of York, Pennsylvania, have calculated the costs of an OTP plant would be one hundred sixty-six dollars per kilowatt;Professor Clarence Zener of Carnegie-Mellon University in Pittsbugh, Pennsylvania, states OTP could cost as little as ninety-six dollars per kilowatt, if plants were mass produced in modular sections; and professor Heronemus (also active in OTP research) computed costs at four hundred dollars per kilowatt, including electrical transmission lines.[33]

OTP costing even two or three times the NSF estimate might still be a valuable power source. OTP would permit the conservation of fossil fuels and would allow us to achieve energy self-sufficiency by stopping the importation of foreign oil, while forgoing total dependence on nuclear power for expanding our generating capacity.

The basic technology for OTP has been around for almost half a century since French scientist Georges Claude demonstrated a twenty-two kilowatt ocean thermal plant at the Bay of Matanzas in Cuba in 1929. Among the significant technological concerns remaining today are those pertaining to the "sophisticated plumbing" needed for the deployment of OTP. According to the *Solar Energy Program Report* and other sources, the engineering problems that must be met before commercial OTP plants can be built require only "relatively low-level technology." The plumbing and hydraulic challanges include production and installation of the enormous deep-water intake pipes, turbine diffuser design, and pump design. The plant may also be susceptible to biofouling (growth of marine organisms) on heat exchange surfaces far below the surface (reducing heat exchanger efficiency), and to corrosion and scale problems resulting from the effects of salt water on the piping.

OTP proponents believe that these engineering problems and other potential objections to the OTP systems can be readily resolved. Although pipes of a thousand feet or more in length and forty feet or more in diameter may be needed for OTP plants, total pipe lengths required for a hundred-megawatt plant are actually comparable to those used in a conventional hydroelectric plant, according to the Andersons.[34] These OTP pipes could be constructed in shipyards using technology developed for the construction of large sewage outfall pipes, or the modular sections could be assembled at the plant site as the pipe is submerged.

Some analysts of OTP proposals have wondered whether pumping cold water up from the ocean depths would cause the cold water to mix with warmer layers above it, thereby defeating the whole purpose of plumbing the sea's nether reaches. Engineers versed in OTP lore, however, believe that with properly designed pipes, the water will scarcely mix. The pumping process itself is not expected to render the operation of the plant uneconomic in an energy sense as no more than 10 percent of the total generating power of the plant should be needed to energize the pumps, primarily because the water never has to be lifted out of itself.

To maintain a constant supply of warm water, the plant may have to be anchored in the Gulf Stream, which also may prove to be a difficult engineering problem. Alternatively, the plant could be designed to propel itself, so as to maintain a steady supply of warm surface water; the propulsion is expected to use 5 percent or less of the plant's power output. Similar self-propulsion methods are currently used to move oil drilling rigs.

Because large OTP systems have prospectively been sited far from shore and in water depths that might make electrical transmission by undersea powerline difficult, it may be advantageous to use the electricity from OTP systems to generate hydrogen from seawater. The resulting hydrogen could be cooled and compressed to a liquid state and shipped to shore by tanker, or the hydrogen could be piped ashore for use.

Shore-based OTP systems — using sea water piped into the plant as the plant's working fluid — could also be designed to produce fresh water for use in arid regions. The plant's condensed (hence desalinated) water would simply have to be collected.

The shore-based OTP plant could also be used as the basis for a thriving mariculture system.[35] Nutrient-rich cold water from the ocean depths could be piped into shallow ponds after use in the plant's condenser. In the ponds, it would be warmed by the sun and would nourish an abundance of algal life, the key to much more complex food chains. In successive ponds, scallops, clams, and oysters could be raised on the algae; culled young mollusks could be fed to lobsters and crabs.*

As with any energy technology, environmental effects are sure to occur, and in the case of the OTP system, they have not yet been carefully studied. These effects seem to be of serious concern primarily if large numbers of OTP plants are deployed. Then huge volumes of water from the depths would be mixed with warmer surface- and middle-strata waters as the cold waters were discharged from the plants. In the Gulf Stream, it has been theorized that a one degree change in temperature might cause a change in the direction and flow of the stream and in bordering terrestrial climate

OTP plants could be constructed by 1985, according to the *Solar Energy Program Report*, provided that an accelerated sea thermal development program is pursued by the federal government accompanied by incentives to stimulate the development of a commercial OTP plant industry. Sea thermal power seems attractive enough as a source of energy to be energetically pursued by ERDA. It deserves a prominent place in any serious survey of plausible alternative energies.

---

*OTP mariculture is being studied by scientists from the Lamont-Doherty Geological Observatory of Columbia University at a test station on St. Croix, Virgin Islands.[36]

TIDAL AND WAVE POWER

Tidal energy, on casual inspection, may seem like another major unheralded energy option for the U.S. But unlike the other alternatives considered so far, the energy recoverable from tides is insubstantial compared to current levels of U.S. energy use.

The gravitational forces of the moon and other solar bodies acting on the earth create the cyclical ocean tides. At each ebb and flow, great masses of water are raised and lowered. This requires enormous amounts of energy because each cubic foot of water weighs more than sixty-two pounds and billions of cubic feet are set in motion with each tide, yet the amount of tidal energy that can be extracted from the tides is small unless a tidal plant is built in an extremely advantageous location.

A tidal plant consists of a dam equipped with flood gates, built across the mouth of a bay. Water can be allowed to flow in through the gates from the sea while the tide is rising, and then can be trapped behind the dam by closing the gates at high tide. When the water on the sea side of the dam ebbs away, the trapped water can be released by opening the gates. As the water descends to the seaward side of the dam, now at low tide, the water is passed through electric turbines that create power, as in a hydroelectric plant.

The electricity available from the plant depends on the volume of the water behind the dam and on the height of the tides (the tidal range). Numerous engineering calculations have established that only broad bays with narrow mouths and large tidal ranges are likely to provide economically competitive electricity. Few sites meet the criteria sufficiently well to repay the energy and capital that must be expended to build tidal power plants, and only one of these sites, Passamquoddy Bay, is found on the U.S. coast. Moreover, it is expensive to build dams in flowing salt water that are strong enough structurally to withstand the sea's constant battering.

It can and has been done, however. The world's only major tidal power plant operates in an estuary at La Rance, France, where the tidal range is twenty to forty-seven feet and the estuary's mouth is relatively narrow (twenty-three hundred feet). This made it possible to complete the plant in 1966 for about four hundred dollars a kilowatt.[37]

A total of perhaps fifteen such advantageous tidal power plant locations exist throughout the world, and as the price of fossil fuels continues to climb, those sites will become more attractive.

Another stratagem for producing electricity from the ocean is to use the horizontal energy of wave trains. Many possible techniques for doing so are currently under consideration by scientists, but these projects are only in very early stages of development. One plan now being studied at the Technion Institute in Israel is the use of a pulley system that would connect vertical concrete pillars sunk in the seabed one hundred fifty to two hundred yards offshore to pulleys on a shore-based generator. Energy from transverse motion of the waves would be delivered to the generator by umbrella-shaped cones fastened to a plastic drive belt on the pulley. With the open side of the "umbrella cone facing seaward, the cone would act like a sail, propelled along underwater by wave energy." On shore, the belt would turn the generator pulley.[38] The feasibility of this and other wave energy studies remains to be demonstrated.

FUSION POWER

Even with good fortune, [fusion] will surely not be generally available before A.D. 2000.

> Professor of Engineering David J. Rose,
> M.I.T., and others

It already appears that a working fusion reactor of the Tokamak variety will have to be large and costly . . . [there is] continuing uncertainty as to whether or not fusion would become an economic energy source.

> Walter Sullivan, *New York Times*
> March 22, 1976

One day, fusion may provide the world with cheap electrical energy, but that day is still decades away. Fundamental research on fusion remains to be done, for fusion is in a far more rudimentary stage of development than either solar electric power or nuclear fission. Fusion's capacity to produce net energy has not even been scientifically demonstrated yet, and important problems remain before this can be achieved. In addition to these scientific challenges, numerous technical, engineering, economic, and environmental uncertainties exist today about fusion.

As in fission, the ancient alchemists' dream of transmuting the elements occurs. But unlike fission, in which heavy elements are split into lighter ones, fusion joins two light atomic nuclei to produce heavier atoms. Ordinarily, the nuclei of atoms are surrounded by negatively charged electrons and trhe atoms resist any attempts to push them together hard enough to make them change thier basic physical nature. But scientists experimenting with the world's first high energy particle accelerators in the 1930s began to wonder whether this resistance could not be overcome by forcing atoms together at high energies. They had recently deepened their understanding of how this produces the energies of stars, such as the sun, and they were eager to induce fusion on earth. A group of physicists, led by Dr. Edward Teller, solved this problem by

Los Alamos Scientific Laboratory

Figure 12-5. Experimental apparatus for controlled fusion research.

creating another: the hydrogen bomb. Teller's group ingeniously used the explosion of a fission bomb to force hydrogen nuclei together and produce fusion. The fission bomb exploited the inertia of the hydrogen mass as a containment device by heating the hydrogen so fast that fusion took place before heat-induced expansion could occur.

To derive electrical power from fusion, the reaction must be controlled to proceed more slowly than in a bomb, so as to provide energy for an extended time. A gas such as hydrogen can be heated to the point that it dissociates into a positively charged proton and a negatively charged electron. Thus ionized, the gas is known as "plasma." Essentially, all fusion methods under study today are attempts to contain plasma under intense pressure at high temperatures in a relatively pure state so that the fusion of atomic nuclei can occur, resulting in the release of huge amounts of energy. Simultaneously, the plasma must be held in a hot enough, small enough space — long enough for fusion to pay back the energy spent in containing the plasma. Both current major approaches to fusion today — magnetic confinement fusion and laser fusion — are simply different means to this same goal.

MAGNETIC BOTTLES

Hydrogen and helium isotopes and the element helium itself are suitable as fusion fuels. The essential ingredient in all the fusion formulae under consideration is deuterium. This isotope of hydrogen has a neutron as well as a proton in its nucleus. Tritium, also a hydrogen isotope and another important fuel, has a proton and two neutrons in its nucleus. Once a deuterium-tritium fuel mixture gets hot enough for its electrons to strip away, the positively charged nuclei of deuterium and tritium will repel each other. To overcome the repulsion, the plasma must be raised to an estimated 100 million degrees Centigrade.[39] This so energizes the nuclei that they fly about fast enough to collide with one another and fuse. Fusion will continue only if the plasma is subjected to such intense compression that the probability of the necessary collisions is high enough to insure their occurrence with the required frequency. As compression of plasma increases, its density is increased hundreds of times.

For fusion to produce more energy than its induction requires, physicists have computed that the product of the plasma density (in particles per cubic centimeter) and its confinement time in seconds must exceed 10^{14} (i.e., a hundred trillion). This physical limit is known as the Lawson criterion, and its achievement is considered a demonstration of fusion's scientific feasibility. As of 1975, fusion experiments were at least ten-fold below the feasibility criterion.[40]

Physicist Hans Bethe has stated that if the Lawson criterion can be met in a laboratory by 1982, this will correspond to Fermi's proof of the feasibility of the nuclear fission chain reaction in 1942.[41] Dr. Bethe asserted that if our experience with fission is any guide, another twenty years beyond the proof of feasibility will be required before commercial implementation of fusion occurs and ten more years would be needed before fusion could make any sizable contribution to U.S. electric power supply. An additional ten years after that would be needed, Dr. Bethe claimed, before fusion would contribute 10 percent of our power. Even these pessimistic projections that fusion will require forty years to become commercially significant *assume* that eventually it will be economically as well as scientifically feasible *and* preferable to nonradioactive alternative modes of energy generation. Nonetheless, some scientific proponents of a crash fusion-development program believe that commercial fusion could be attained much sooner than Dr. Bethe states.

Once scientific demonstration of fusion has been accomplished, difficult engineering and materials-selection problems remain. To simultaneously superheat and compress plasma, a novel containment method had to be invented. At the temperatures necessary for fusion, the plasma is so hot it would melt any physical container. Moreover, any contact between the container wall and the plasma would reduce the plasma temperature, bringing fusion to a halt.

Most fusion research to date has focused on containing the plasma by magnetic fields inside a vacuum. The machines designed to perform this function have strange names, such as the magnetic mirror reactor, the Reference Theta-Pinch Reactor, the Minimum-B Magnetic Mirror with Yin-Yang Coils, and the Tokamak toroidal reactor. All these designs subject plasma in a large vacuum tube to a tremendous electromagnetic field exerted by magnets surrounding the tube. Many complex engineering problems have been encountered in magnetic confinement fusion. Temperatures of 100 million degrees centigrade must be induced in the plasma while, about two yards away, the temperatures of the superconducting electromagnets must be near absolute zero (-273.16°C). Surrounding the vacuum but insulating the coils of the electromagnet is a "wall" of molten lithium at about one thousand degrees centigrade. The lithium "blanket" is needed to trap high-energy neutrons released by fusion and keep them from the magnet. The impact of the neutrons on the lithium produces both radioactive tritium and heat. The heat can be transferred by a heat exchanger to a steam generator and turbine for producing electricity.

Several possible reactions can release fusion energy. Deuterium can be fused with deuterium, producing a proton, an atom of tritium, and energy, or a neutron, an atom of helium-3, plus energy. The newly produced tritium also can be fused with deuterium, producing a neutron, helium and energy. Because the deuterium-tritium reaction requires the lowest temperature, it is likely to be the first to be fully developed.

The great advantage of fusion is that deuterium fuel is virtually inexhaustible and the energies released from its fusion are enormous. Deuterium is found in sea water and can be recovered cheaply in a small deuterium plant. The product energies released in fusion range from twenty-two thousand kilowatt hours per gram of reacting nuclei in the deuterium-deuterium reaction to ninety-eight thousand kilowatt hours per gram of nuclei in deuterium-helium-3 reactions.[42]* Moreover, the byproducts of the fusion reaction, helium and tritium, are themselves fusion fuel.

Advanced fusion reactors may be able to fuse deuterium and helium-3. Their fusion products are all elecrically charged and can be withdrawn as current directly from the reactor after fusion occurs without the necssity of a steam cycle. Such a plant has a higher potential efficiency than the steam-cycle fusion reactor.

ENVIRONMENTAL IMPACT

Contrary to popular assumption, fusion reactors would pose environmental problems, and the experience base on which to assess their seriousness is still negligible. Fusion reactors will contain a total of about 100 million curies of tritium.[43] This light, radioactive gas has a half-life of twelve years and an uncanny ability to diffuse through hot metals at the temperatures in a fusion reactor. Not only can tritium be a hazard if inhaled in gaseous form, but because of its similarity to ordinary hydrogen, its chemical behavior is the same and it can replace hydrogen in water. Conse-

---

*Readers who remember the high school chemistry experiment in which hydrogen and oxygen combine in a loud explosion to form water and energy may be interested to know that the least energy-producing fusion reaction of deuterium produces more than half a million times as much energy as the hydrogen-oxygen reaction.

quently, the human body has no way to discriminate between the two elements and readily incorporates radioactive water. One mitigating feature of the tritium-deuterium fuel cycle is that tritium for a fusion reactor could be both produced and consumed in the same locale, making transportation unnecessary, in contrast to current practice in the nuclear fuel cycle.[44] The safety of the reactor would still be contingent, among other things, on how well the tritium is contained at the fusion facility.

The intense barrage of neutrons produced in a fusion reactor creates radioactive materials in the reactor's metal walls, which may be made of niobium or vanadium alloy. Fusion reactors using vanadium alloy may have biological health effects an order of magnitude or two less than fission reactors. Neutron activation products in a fusion reactor's structure will go on producing heat from radioactive decay long after the reactor is shut down, similar to the afterheat of a fission reactor. The intensity of this afterheat for a niobium fusion reactor shortly after shutdown is thought to be comparable to the afterheat in a fission reactor, but since cessation of cooling will not cause radioactive fuel bundles to melt, as in fission, the afterheat problem is far less serious for fusion reactors.[45]

Fusion reactors may have other operating hazards which are not shared by ordinary fission reactors. Liquid lithium metal surrounding the fusion reactor core is highly flammable and, if it comes into contact with water, it could explode. The large magnetic fields of fusion reactors may also be dangerous; the energy in the magnetic system of a toroidal (doughnut-shaped) Tokomak reactor is equal to a lightning bolt's.

A new hybrid fusion/fission reactor has been proposed by scientists to use the high-energy neutrons created in fusion reactors for the production of plutonium 239 from uranium 238 that would be contained in a blanket around the reactor core. Soviet scientists already are planning a reactor of this type, the Tokomak-20. It will produce large amounts of radioactivity and the plutonium may add to our already severe nuclear weapon proliferation problems.

NEW DEVELOPMENTS IN MAGNETIC CONFINEMENT FUSION

Erda officials announced in November 1975 that scientists at the Francis Bitter National Magnet Laboratory of Massachusets Institute of Technology, using a Tokomak fusion reactor, had achieved a plasma density-time product of 10^{13} particles per cubic centimeter at temperatures of ten million degrees centigrade.[46] That product is five times higher than any previously attained. Yet temperatures ten times higher would be needed to meet the Lawson criterion.

Laser-initiated fusion is off to a much later start than magnetic confinement fusion, but it has been advancing with great speed. In laser fusion, plasma is held together by inertial confinement as in a hydrogen bomb explosion.

The fuel is compressed into tiny glass pellets for injection into the reactor where pulses of laser energy blast away at the pellets from several angles and compress the fuel more than a thousand times. This heats the fuel so rapidly that the fuel does not have time to overcome its inertia and expand. Physicists believe that ten times greater compression will be necessary, however, before the energy released by the fusion even equals the enormous power needed to operate the lasers. For *net* energy to be produced, lasers with energies comparable to those that would be needed to lift four tons in a billionth of a second are required.[47]

One of the major fusion research centers in the U.S., the Lawrence Laboratory at Livermore, California, will be building twenty neodymium glass lasers, each one hundred forty feet in length.[48] Each of the lasers is designed to deliver high-energy light pulses. One pulse delivers six trillion watts of power — a rate comparable to the entire average power output of the U.S. But the pulse lasts for only one hundred fifty trillionths of a second. Scientists at Livermore are hoping that by concentrating twenty such laser beams on a fuel pellet, they will come closer to attaining electrical power in the form of net energy from fusion. Dr. A. Carl Haussman, director of Livermore's fusion effort, states, however, that lasers a hundred times more efficient than the neodymium devices will be needed before laser fusion power plants can be used.

The successful development of fusion power will require a

costly research effort that is unlikely to produce commercial fusion for at least another twenty-five years. In the long run, *if* it can be proven safe and economic, as its proponents believe, then it will be a valuable addition to our energy supply. But safer, less expensive and already feasible solar alternatives are available *right now* and priority should be given to their development!

OIL SHALES

Tremendous amounts of energy are locked in U.S. oil shale deposits, but shale oil today still costs much more than natural crude oil. Experts estimate that shale oil would cost about twenty-two dollars a barrel today (in 1975 dollars), if produced by one well known process (Tosco). In a sense, this is fortunate, because the environmental problems af recovering shale oil and the energy needed for the recovery processes are formidable.

Oil shale deposits currently hold more energy than all U.S. oil and gas reserves combined. According to the Ford Foundation report, *A Time To Choose*, the U.S. has enough high-grade shales (over twenty-five *gallons* of oil per ton) to supply an amount of oil equal to 1973 U.S. oil consumption for a hundred years.[49]

Oil shale is a fine-grained sedimentary rock (laminated marlstone) containing kerogen, a tarlike hydrocarbon. When the shale is heated, a gas and a heavy oil are produced. The oil is then "upgraded" to produce a synthetic crude oil that is equivalent in energy value to high-grade crude oil.

The production of shale oil is a highly energy-intensive process because of the low oil-to-volume ratio found even in high-grade oil shale, and because the chemical retorting and refining processes require the addition of substantial amounts of heat energy.*

Shale oil production also creates a vast waste disposal problem: For every barrel of shale oil produced, up to one and a half tons of rock may have to be processed (depending on the ore grade). Because of this high input-to-output ratio, a fifty-thousand-barrel-per-day retorting operation using 35 gallons-per-ton ore would produce

---

*In the opinion of Professor G. Tyler Miller of St. Andrews Presbyterian College, "At best [shale oil] may produce only a slight gain in net energy at a high cost."[50]

sixty thousand tons of spent shale daily — enough to cover twenty acres of land a foot deep.[51] Spent shale has a greater volume than the more compact shale rock in the ground, so it cannot just be put back in the mines from which it comes. Entire canyons and valleys would have to be filled with the waste material.[52]

Commercial underground mining for shale may turn out to be roughly twice as expensive as surface mining, and the temptation to strip mine it might prove irresistible to the energy combines. Already, two of the four federal oil-shale tracts so far leased for development under a prototype leasing arrangement are slated for strip mining. And because oil shale beds can be more than a hundred feet thick, surface oil-shale mines for exploiting the massive deposits will be deeper and larger than surface coal mines, instead resembling huge open-pit copper mines.

In addition, 90 percent of the U.S. oil shale resources are in the Green River Formation, a geological entity underlying twenty-five thousand acres of land in Western Colorado, Utah, and Wyoming; seventeen thousand acres of which are thought to contain commercially recoverable oil shales. The region is arid-to-semi-arid with an extremely low population density and scarce water supplies.

Oil shale complexes are likely to require huge quantities of water. Crushing, retorting, upgrading, and reclaiming shale rock all demand water and contaminate it with a variety of pollutants. If surface shale retorting is employed, a million-barrel-a-day shale-oil industry would need 121,000 to 189,000 acre-feet of water annually just to reclaim the solid wastes.[53] That is about 10 percent of all the water now used in the Colorado part of the Upper Colorado River Basin. It is doubtful that this much water could be procured locally without severely infringing on the water rights of prior users in the area and downstream, where the water is used for farms, ranches, and other industry. Projects in which large amounts of water were removed from other regions for an oil-shale industry would have alarming environmental consequences, too.

Processes for retorting the shale deep within the ground (*in situ* retorting), without necessitating mining, are still in early stages of technical development. The shale would be fractured below ground, by one of various techniques, and it would be heated to decompose it into gas and oil so the oil could be withdrawn by wells. If *in situ* retorting can be done commercially, solid wastes would be almost absent or far less than from surface mining and retorting. Water requirements for an oil-shale complex based on *in situ* retorting would probably be much less than for a surface retorting operation because reclamation requirements would be lower. Nonetheless, environmental consequences in terms of water use, pollution, and air pollution would be serious. For one *in situ* process (BuMines), air pollution from each trillion Btu of oil shale retorted would contain two hundred sixty-two tons of sulfur dioxide, eleven tons of hydrocarbons, and six tons of particulates. Furthermore, a power plant would be needed at the complex to supply energy for the retorting.

In sum, severe environmental impacts appear likely to occur with oil-shale production. Unless we are willing to see the wholesale destruction of vegetation, animal life, and clean water in one of America's precious but dwindling undeveloped regions where majestic wildlife still abounds today, oil-shale development should not begin.

GARBAGE POWER

Garbage contains dry organic solids that can be used, after processing, for the production of power, heat, or fuels. If all U.S. refuse could have been collected and economically utilized in 1971, its energy content would have been equivalent to 13 percent of our total energy consumption or to 44 percent of our electricity use. Some refuse resources are too diffuse to collect, however, without expending an inordinate amount of money or energy, but the amounts of readily collectible refuse in 1971 would have been enough to produce 2 percent of U.S. energy needs in that year.[54] If converted to electricity, this would have been more than the power produced then by nuclear plants.

The useful components of refuse include agricultural crop residues, manure, municipal refuse, sewage sludge, and timber refuse from logging and lumber milling. Once collected, organic waste must be shredded, sorted, and may be subjected to further processing, such as drying or magnetic separation.

Each year, Americans throw away about 134 million tons of refuse. The burnable organic material in the waste has as much energy as $1.5 billion worth of imported oil, or 70 percent of the annual output from the Alaskan oil fields. The wastes also contain iron, steel, and aluminum, whose original refinement required large amounts of energy. But by reclaiming these materials, much energy can be saved. It takes only 5 percent of the energy to retrieve a ton of aluminum from scrap, for example, as to refine the metal from bauxite ore.[55]

Solid waste recovery and conversion of refuse to low-sulfur fuels operate in tandem. First, recyclable materials are removed from refuse, then the organics remaining are used to produce low-sulfur fuel oil by pyrolysis. In pyrolysis, organics are heated in the absence of oxygen, resulting in decomposition to gas, char, and the fuel oil. Less sophisticated fuel-recovery plants simply prepare shredded organic matter in scraps or briquets that can be burned together with coal in the furnace of a coal power plant.

Organic wastes in garbage also can be converted to oil (with a heating value of fifteen thousand Btu per pound) by a process called hydrogenation. Water and carbon monoxide are added to the organic material at high temperature and pressure and allowed to react. Hydrogen and carbon dioxide are produced causing the hydrogenation of the organics to oil. This technology is in the pilot-plant stage and its costs are uncertain.

According to the *Sierra Club Bulletin*, at least eighteen cities are now designing "energy-recovery facilities that will use solid waste as fuel."[56]

Plans to use garbage power seem to be most advanced in Connecticut where the Connecticut Resources Recovery Authority has launched a $290 million plan to collect all the state's domestic and commercial refuse for conversion into electricity and valuable scrap.[57] Refuse from one hundred sixty-nine towns will be collected into ten regional centers by 1980, instead of at hundreds of polluting incinerators and dumps. The state's first two plants are due to begin operating by mid-1976. They are designed to process thirty-six hundred tons of garbage a day, producing fuel at a profit to supply 10 percent of the electricity in their areas, Bridgeport and Hartford. This system is expected to save Connecticut taxpayers $100 million just five years after the project is complete, with large reductions in air pollution and in land needed for solid-waste dumping. Adoption of the Connecticut system in other states could have a significant short-term impact on U.S. electrical supply, substantially weakening any "rationale" for adding uranium fission plants.

PLANT POWER

Plant biomass or living matter can be dried and substituted for a limited amount of rapidly depleting fossil fuels, but in a world fraught with catastrophic food shortages, it would be irresponsible to devote valuable land capable of food production solely to fuel generation, when other renewable energy resources are available. Unlike solar-power technology, which can be operated in desert regions, fuel crop production, except wood, requires arable land. High crop yields also usually require irrigation, fertilizers, and substantial energy inputs.

Large land areas are needed because plants are generally inefficient converters of solar energy to recoverable heat. Efficiencies of 3 percent during a growing season are all that can be expected, and the year-round average is about 1 percent. Consequently, two hundred fifty to five hundred square miles of land would be needed to fuel a 1000 MWe steam-cycle power plant, based on optimal dry-weight biomass yields of ten to thirty tons per acre annually. The most efficient high yield crops (in tons per acre-per year) are algae (fifteen to thirty tons); cultivated forest (twenty tons); eucalyptus (eight to twenty-five tons).

According to one economic assessment, assuming seventy five hundred Btu are available from each pound of biomass product, the cost per million Btu from biomass would range from $0.80-$1.20, compared to $0.79 for coal and $0.87 for oil.[58] Therefore, the lowest cost-estimate biomass fuel could compete with coal at current prices and could compete very favorably with more expensive nuclear power. Biomass as a cash crop is roughly equivalent in value to the dollar yield per acre of wheat. Synthetic algae culture in ponds (versus oceanic culture) has been found by one researcher to be too costly to make the use of the algae economic as fuel.[59]

The dried stalks and husks of grain crops such as sorghum, corn, and wheat have approximately two thirds the heating value per pound as coal. They can be used with coal in power plants to lower sulfur emissions or in the production of liquid fuels, such as ethanol or gasoline substitutes. It has been suggested by some experts that enough ethanol or other gasoline substitute could be produced in this way to replace all the oil now imported from abroad. The cellulose in the agricultural residues would be enzymatically converted to glucose which would then be fermented to produce ethanol. Instead, most agricultural residues are wasted today, although up to two thirds could be removed while still leaving enough as ground cover to protect the soil from erosion.

Methane can also be generated in anaerobic digesters from organic wastes such as sewage sludge and agricultural manures.* A digestor is a sealed compartment where organic wastes are decomposed by bacteria in the absence of oxygen. The process yields methane gas and a valuable odorless organic fertilizer that is high in nitrogen. Scientist Hinrich Bohn, writing in *Environment* (December 1971), has concluded that the digestible organic waste now available in the U.S. would supply 1.5 times the amount of natural gas consumed.

The use of land, water, and fertilizer for biomass production becomes less extravagant when a useful food or industrial crop in addition to the raw biomass is produced. Sunflower fields, for example, can yield oil and seeds, in addition to stalks suitable for combustion or pyrolysis.

Oil and auto industry researchers have expressed interest in growing grain for conversion by fermentation to methyl and ethyl alcohol for use as fuel, but it is difficult to justify growing grain exclusively for fuel in a world experiencing a chronic food shortage.

COAL

Ninety-five percent of U.S. energy in 1973 came from fossil fuels, and about 18 percent was from coal.[60] Critics of coal are concerned about three major aspects of its use: occupational health problems affecting workers in the coal industry; public health problems affecting the population at large; and environmental impacts of coal use. Each of these categories has, in turn, three principal dimensions: the mining/transportation stage; the beneficiation/refinement stage; and the combustion-at-power-plant-site stage. To consider the subject of coal power in a comprehensive way, all these factors — a ninefold matrix of variables — need to be assessed. Three major mining

*Methane is the main constituent of natural gas.

techniques, five distinct coal conversion technologies, and their various commercial variants further complicate the analysis. The brief section that follows merely outlines several conclusions about the safety of coal-generated power.

Some fission supporters have muddled the complex issue of coal safety versus nuclear safety by comparing the impacts of an archaic, dirty coal technology without pollution controls to those of the most modern fission plants, in which advanced pollution-control devices are used to minimize routine radioactive releases. A proper comparison of fission to coal combustion requires that both be compared at similar stages of technological development.

Another analytical pitfall also needs to be skirted. This is the popular but mistaken notion that the U.S. either must choose large-scale increases in coal combustion or large-scale deployment of fission plants. *In fact, both these options can be foregone entirely.*

As demonstrated in Chapters 10 and 11, conservation, redesign of utility rate structures, and both home and commercial uses of solar power can be substituted for nuclear fission in the short run, strictly limiting the need for new power plants of all kinds. In the medium term (ten to fifteen years), solar power, energy conservation, and gasified coal can be accompanied by the expanded use of geothermal power and by phasing in large-scale solar power plants using direct as well as indirect solar technologies. As this occurs, coal plants might be phased out, without ever having to phase large-scale nuclear fission power in. Thus no short run choice between coal and nuclear power is necessary!

COAL TODAY

The coal industry today is an occupational and public health hazard and an environmental tragedy.

Deep coal mining as now practiced in most U.S. mines is one of the most dangerous occupations. Tens of thousands of coal miners have been disabled by black lung disease (coal workers pneumoconiosis). But if present standards for mine dust control are strictly enforced, experts believe black lung disease can be avoided. Accidents and injuries also can be greatly reduced. People in the coal mining and processing industry on the average suffered twice as many days of disability in 1972-1973 as those in other high-risk occupations such as construction or metal mining and milling.

But a wide range of injury rates was apparent within the coal industry: some U.S. coal mines had lower injury rates than for professions like teaching and government service, and the injury rate variation from safe to unsafe mines ranged more than twenty-fold in 1972-1973. Thus tremendous improvements in deep mine safety are possible. This conclusion is supported by fatal-injury-rate data from the United Kingdom showing that British mine fatalities were a half to a quarter of U.S. rates in the early 1970s. The U.S. coal industry, according to the Ford Foundation energy project (*A Time To Choose*), could be made at least as safe as other industrial operations in the U.S.[61]

The environmental effects of deep coal mining are often serious. According to Dr. Allen Hammond, "It is responsible for a long string of environmental insults, including acid mine drainage, erosion, silting of streams, leaching of pollutants from accumulated wastes, fires in mine and waste piles, and subsidence of land over mined-out cavities."[62] Some of these problems can be significantly mitigated by careful mine management.

About half the coal in the U.S. comes from surface mines of the area or contour type where coal is stripped from seams or beds near the surface after excavation of the overburden. Serious solid-waste and water-pollution problems result from strip mining, and land may be permanently ruined for many future uses. Erosion and acid mine drainage can damage streams and other surface water sources. Valuable underground water supplies normally capped by coal seams, can also become ruined by strip mining of the coal seal.

A typical contour strip mine producing ten thousand tons of coal a day will produce one hundred twenty thousand tons of solid waste daily, an amount equivalent in weight to the municipal refuse from a city of forty-eight million people.[63] To date, not even half the areas strip-mined in the U.S. have been "adequately restored and a land area the size of Rhode Island and Delaware has been severly scarred."[64]

Reclamation of strip-mined land requires reshaping the land to restore its surface, replacement of topsoil, fertilization, revegetation and supervision of the land until it returns to productive use. In areas where reclamation is possible, it needs to be planned at the time mining is initiated and then care of the land must continue for years after mining ceases. According to the Ford Foundation, "No such efforts have as yet been undertaken in this country."[65]

Even where attempted, reclamation is often imperfect or impossible. Replaced material has to be contoured by bulldozers or other machines to counter erosion, creating a monotonous, homogeneous, artificially graded region in place of natural land contours and vegetation. Sometimes artificial terraces and drainage ditches have to be installed. In areas where the land is arid to semi-arid, as in much of the West, reclamation is not possible because the soil is too dry to hold moisture well, and surplus water in sufficient quantities is unavailable.

Nonetheless, the coal mining industry tends to employ strip mining because there are fewer safety problems in surface mining and, more importantly, because stripping generally costs only half as much as deep mining, although precise costs depend on mining technology used, the nature of the deposits being mined, and production volumes.

Fortunately for the U.S., the U.S. can abandon strip mining within fifteen years, and obtain all its coal from deep mining, according to John McCormack, coal expert for the Environmental Policy Center in Washington, D.C.

CLEANER COAL

Today's coal combustion technologies produce or contribute to a number of very serious environmental problems. One difficulty, apparently shared by all fossil-fuel technologies, is the addition of carbon dioxide to the atmosphere. By absorbing heat reflected from the earth towards the sky, the carbon dioxide gradually increases the temperature of the earth and could, by shifting the earth's temperature as little as one or two degrees centigrade, produce major and undesirable climate changes.

Neither solar nor nuclear technologies present this drawback. In addition to carbon dioxide, coal plants produce unwanted heat, although because of their greater thermal efficiency, they produce proportionally less than nuclear fission plants. A primary problem with coal technologies is the air pollution they create. Sulfur found in coal accounts for much of the pollution. The sulfur forms sulfur oxides that are converted gradually to sulfuric acid and sulfates, which are more dangerous than sulfur oxide. Some of this coal pollution is emitted in the form of tiny submicron-sized*particles that are especially harmful to human health, because they are small enough to get past the filtration systems of the lung and can penetrate deeply into the bronchioles, carrying sulfates and trace quantities of other toxic substances released by combustion. The particles' size also makes them difficult to remove by pollution control devices. Although the rate of sulfur removal is high at a modern coal plant where scrubbers and electrostatic precipitators are used, environmental problems remain because as yet neither of these pollution-control technologies is particularly effective in removing small particles.

Some coal plants are equipped with flue-gas scrubbing devices. In a scrubber, exhaust gases are passed through streams of water treated with sulfur acceptors to absorb sulfur. Of various scrubbing processes, only lime and limestone scrubbing have been shown commercially successful.[66] Lime scrubbing can remove up to 90 percent of stack-gas sulfur into a lime slurry (water lime suspension).

More advanced techniques for sulfur removal are also available. Coal can be burned in a fluidized bed system in which the combustion air reaching the coal has been passed through several feet of granular lime, coal ash, or dolomite. Particles of the granulate are fluidized by the air and combine with sulfur. Although not yet commercially available, fluidized bed combustion will increase the efficiency of coal plants and it probably will be able to remove 95 percent of the coal's sulfur.

COAL GASIFICATION

Coal can be burned more cleanly and efficiently if first converted to a gas. A growing market for synthetic coal gas exists because U.S. natural gas production peaked in 1973 and has been restricted by the oil companies since. Natural gas consumption is growing at roughly two and a half times the rate of new annual discoveries and present reserves will last only another decade, at current consumption rates.†

---

*A micron equals a millionth of a meter.

†These figures come from the industry group, the American Gas Association. Independent assessments are not readily available.

Fortunately, coal can be converted to gas of low, medium, or high Btu value, and can also be liquefied to an oil. Commercial gasification technology has existed at least since World War II. Low Btu gas conversion is the simplest and most economic of the gasification/conversion technologies. Coal is heated so it decomposes, releasing a gas which is a mixture of carbon monoxide, carbon dioxide, hydrogen, nitrogen, and methane. The main advantage of gasification is that sulfur can be removed during the process before combustion occurs. Operating at much higher temperatures than conventional coal combustion, gasification yields ash and sulfur in the form of hydrogen sulfide gas, which is more easily removed from stack gas than the sulfur dioxide of ordinary coal-fired plants; moreover, tars and condensable hydrocarbons can be eliminated totally. Coal gasifiers can therefore use as feedstock high-sulfur coal which creates too much pollution to be burned in coal plants without emission controls. Current gasification technology can produce gas usable for firing power plants and for chemical industries, like those producing ammonia and methanol. The latter industries can use gas with a heating value of three hundred Btu per cubic foot. This gas is not suitable for mixing with the higher quality, thousand Btu-per-cubic-foot natural gas flowing through major pipelines.

Using one low-Btu gasification process (Lurgi), 99.7 percent of the fuel's sulfur can be removed. The resulting coal gas can be used in a gas turbine combined-cycle power plant. Gas is burned in the plant and the hot exhaust gases first spin a turbine and are then fed to an unfired boiler. There steam or a secondary fluid vapor is used to power a conventional steam-cycle generator. Most such plants are fired with natural gas today, but they will probably be fired by clean-burning coal gas in the future. Because the gas essentially does double duty, efficiencies of combined-cycle plants are much higher than conventional coal plants and several systems are commercially available now (Lurgi, Koppers-Totzek, Winkler, Wellman-Galusha). Several Lurgi combined-cycle plants are already operating, under construction, or on order.

Considerable uncertainties exist concerning future costs of power from low Btu/combined-cycle plants (LBCC), although the range of uncertainties is smaller for LBCC than for future light-water-reactor costs.[67] Despite the LBCC cost uncertainties, further development of combined-cycle plants is going to be a great deal less expensive in research and development costs than breeder reactors, for example.

Although the combustion of coal in LBCC plants looks appealing, its attractiveness is contingent on (1) the rejection of strip mining; (2) the modernization of underground mining conditions; (3) the minimization of occupational health hazards at gasification plant sites; and (4) the avoidance of major environmental problems there.

A preliminary study of synthetic fuel manufacture, including coal gas, liquids, and oil shale, made by the Scientists' Institute for Public Information in New York, has warned that tens of thousands of workers could be exposed to cancer-causing chemicals during synthetic fuel production.* Such occupational health problems must be clearly understood, evaluated, and reckoned with before the U.S. embarks on a major gasification program. Gasification plants may also have serious environmental impacts, especially if they are built in the West. Water requirements are large and could be a potentially serious source of water pollution unless adequate treatment practices are employed.†

The Environmental Policy Center's John McCormack believes there is a danger that, with federal assistance, coal gasification plants with large water requirements might be built next to strip mines in New Mexico and North Dakota where water is scarce, and that this construction would attract a steel industry, resulting in heavy industrialization of the area. If more coal gasification plants are built, a saner policy would be to obtain coal for gasification from deep Eastern coal mines, meeting rigid safety and health standards, in that water is far more abundant in the East than in the West.

---

*"Synthetic Fuels and Cancer," is available from SIPI, 49 East 53rd Street, New York, N.Y. 10022.

†Annual water requirements for coal gasification and combined-cycle systems are about 2.4 million gallons per megawatt, which is about 40 percent of the requirements for a light-water reactor, although more of the water for the reactor can to some extent be recycled, as can portions of the gasification plant water.

CONCLUSION

Coal is a fascinating energy resouce because it is both a challenge and a threat. The mining of coal can be done safely and its combustion in combined-cycle plants can be achieved with public health impacts that are three hundred times less harmful than the emmisions from coal with the best scrubbers available.[68] It is likely that currently available technology can reduce emissions at the gasification plant site to those of oil refineries, albeit at greater cost.[69] Occupational health hazards in the gasification stage should be carefully scrutinized and reduced to acceptable levels, if the technology is to be further employed. Even with all these precautions, coal will add to the earth's ambient heat (through thermal pollution), and also to its burden of carbon dioxide, possibly resulting in global warming. This warming is liable to have unpredictable long term effects on the earth, and therefore can legitimately be compared in seriousness to various long-term problems caused by nuclear fission technology.

Subject to those qualifications, clean-coal technologies could be used to supply interim power needs that cannot be supplied by conservation and cleaner alternatives. Given the environmental penalties which both fission and coal exact from society for their use, it is indeed fortunate that we need not choose to expand greatly our use of either in the near future and, in the long run, we may be able to phase out coal in addition to phasing out uranium fission power.

NOTES

Geothermal

1. Axtmann, Robert C., "Environmental Impact of a Geothermal Power Plant," *Science,* Vol. 187, No. 4179 (March 7, 1975).

2. "New Geothermal Studies," *UC Clip Sheet,* Vol. 50 No. 42 (June 3, 1975).

3. *Ibid.*

4. "U.S. Is Drilling in Montana in Geothermal Energy Test," *International Herald Tribune,* June 17, 1974.

5. Healy, Timothy J., *Energy, Electric Power, and Man* (San Francisco: Boyd & Fraser, 1974).

6. Champion, Dale, "Power from the Earth — Tapping the Geysers," *San Francisco Chronicle*, August 26, 1974.

7. Hartley, Fred L., Open Letter to Employees and Shareholders, Union Oil Co. of California, November 3, 1975.

8. "Geysers' Escaping Gas," *San Fancisco Chronicle*, June 26, 1974.

9. Axtmann, *op. cit.*

10. " 'Slowdown' on Geothermal Land," *San Francisco Chronicle*, October 4, 1974.

11. Alexander, George, " 'Big Oil' Cornering Geothermal Field," *Los Angeles Times*, as reported in the *San Francisco Chronicle*, November 18, 1974.

12. Perlman, David, "Geothermal Stall Charged," *San Francisco Chronicle*, May 21, 1975

Wind

13. Sencenbaugh, James, "Wind Driven Generators," *Energy Primer, Solar, Water, Wind and Biofuels* (Menlo Park, Calif.: Portola Institute, 1974).

14. "Course in Windmills is Offered By a University in New Mexico," *New York Times*, December 21, 1975.

15. Wade, Nicholas, "Windmills: The Resurrection of an Ancient Energy Technology," *Science*, vol. 184, no. 4141, June 7, 1974.

16. Science and Public Policy Program, *Energy Alternatives: A Comparative Analysis* (Norman, Oklahoma: University of Oklahoma, 1975). U.S. G.P.O. 20402. $7.45 Prepared for ERDA, CEQ and various other federal energy agencies.

17. National Science Foundation/National Aeronautics and Space Administration Solar Energy Panel, *An Assessment of Solar Energy as a National Energy Resource* (College Park, Md.: University of Maryland, 1972).

18. Inter-Agency Technical Review Panel, Sub-Panel IX, *Solar Energy Program Report*, December, 1973.

19. Heronemus, William E., "Windpower: A Significant Solar Energy Resource," *Aware*, No. 57, June, 1975.

20. Science and Public Policy Program, *op. cit.*

21. von Hippel, Frank and Robert H. Williams, "Solar Technologies," *Bulletin of the Atomic Scientists,* November 1975.

22. Inglis, David R., "Wind Power As An Alternative to Nuclear," *Congressional Record*, July 14, 1974.

23. Heronemus, William, "The U.S. Energy Crisis: Some Proposed Gentle Solutions," Paper presented to the joint session of American Society of Mechanical Engineers and Institute of Electrical and Electronic Engineers, West Springfield, Mass., January 12, 1972. *Congressional Record,* 92nd Congress, 2nd Session, Vol. 118, No. 17, Park II, pp. 1043-1049.

24. von Hippel and Williams, *op. cit.*

25. Heronemus, William E., "Pollution-Free Energy from Off-Shore Winds," Marine Technology Society, Washington, D.C., September, 1972.

26. Sórensen, Bent, "Energy and Resources: A Plan is Outlined According to Which Solar and Wind Energy Would Supply Denmark's Needs by the Year 2050," *Science,* July 25, 1975.

27. Hughes, W. L., *et. al.,* "Basic Information on the Economic Generation of Energy in Commercial Quantities from Wind," Report ER 74-EE-7, Oklahoma State University, May 21, 1974. Cited in reference 16.

28. Ramakumar, R., H. J. Allison and W. L. Hughes, "Review Paper, Prospects for Tapping Solar Energy on a Large Scale,," *Solar Energy,* v. 16, pp. 107-115 (Great Britain: Pergamon Press, 1974).

29. Heronemus, William, "The U.S. Energy Crisis," *op. cit.*

30. Science and Public Policy Program, *op. cit.*

31. "ERDA and NASA To Dedicate 100 KW Wind Turbine at Sandusky, Ohio," *Information from ERDA,* v. 1, n. 30 (October 22, 1975). "PG&E Taking a Flier on the Wind," *San Francisco Chronicle,* November 26, 1975.

Sea Thermal

32. Inter-Agency Technical Review Panel, *op. cit.*

33. Clark, Wilson, *Energy for Survival* (Garden City, N.Y.: Doubleday and Co., Inc., 1975).

34. Clark, *op. cit.*

35. Othmer, Donald F., and Oswald A. Roels, "Power, Fresh Water, and Food From Cold, Deep Sea Water," *Science,* vol. 182 (October 12, 1973), pp. 121-125.

36. Swann, Mark, "Sea Thermal Power: One Form of Solar Power" (Dublin, Calif.: Committee for Nuclear Responsibility, 1974). P.O. Box 2329, Dublin, Calif. 94566.

Tidal and Wave Power

37. Healy, Timothy J., *Energy, Electric Power, and Man* (San Francisco, Calif.: Boyd and Fraser, 1974).

38. "Getting Electricity from the Sea," *San Francisco Chronicle*, February 27, 1975.

Fusion

39. Hammond, Allen L., William D. Metz, and Thomas H Maugh, *Energy and the Future* (Washington D.C.: American Association for the Advancement of Science, 1973).

40. Sullivan, Walter, "Hydrogen Fusion Studied as New Source of Energy," *New York Times,* October 24, 1975.

41. Bethe, Hans, Professor of Physics, Cornell University, public lecture, University of California, Santa Cruz, March, 1975.

42. Post, R. F., and F. L. Ribe, "Fusion Reactors as Future Energy Sources," *Science,* vol. 186, no. 4162 (November 1, 1974).

43. Hammond, *et. al., op. cit.*

44. Rose, David J. Patrick W. Walsh, Larry L. Leskovjan, "Nuclear Power Vis-A-Vis Its Alternatives, Chiefly Coal," undated unpublished monograph, received January 1976. Dr. Rose, a well-known specialist on nuclear power, is professor of engineering at MIT.

45. Post and Ribe, *op. cit.*

46. *Information from ERDA,* Weekly Announcements, Vol. 1 No. 3 (November 12, 1975).

47. Sullivan, *op. cit.*

48. Petit, Charles, "Potential Electrical Power from Lasers," *San Francisco Chronicle*, September 28, 1974.

Further References on Fusion

Sullivan, Walter, "New Fusion Reactor Likely to Break Even in Fuel Use, *New York Times*, March 22, 1976.

Kammash, Terry, *Fusion Reactor Physics, Principles and Technology* (Ann Arbor, Mich.: Ann Arbor Science Publishers, 1976).

Oil Shale

49. Ford Foundation Energy Policy Project, *A Time To Choose, America's Energy Future* (Cambridge, Mass.: Ballinger Publishing Co., 1974).

50. Miller, G. Tyler, Jr., *Energy and Environment, Four Energy Crises* (Belmont, Calif.: Wadsworth Publishing Co., 1975).

51. Science and Public Policy Program, *Energy Alternatives: A Comparative Analysis* Norman, Okla.: University of Oklahoma, 1975).

52. Miller, *op. cit.*

53. Science and Public Policy Program, *op. cit.*

Garbage Power

54. Science and Public Policy Program, *op. cit.*

55. Rensberger, Boyce, "Coining Trash," *New York Times Magazine*, December 7, 1975.

56. Heffernan, Patrick, "Jobs and the Environment," *Sierra Club Bulletin*, April, 1975

57. Knight, Michael, "Plants to Make Fuel of Garbage," *New York Times*, May 16, 1975.

Plant Power

58. Science and Public Policy Program, *op. cit.*

59. Clark, Wilson, *Energy for Survival* (Garden City, N.Y.: Doubleday and Co., Inc., 1975).

Coal

60. Ford Foundation Energy Policy Project, *A Time To Choose, America's Energy Future* (New York: Ballinger & Co., 1974).

61. *Ibid.*

62. Hammond, Allen L., William D. Metz, and Thomas H. Maugh II, *Energy and the Future* (Washington D.C.: American Association for theAdvancement of Science, 1973).

63. Science and Public Policy Program, *Energy Alternatives: A Comparative Analysis* (Norman, Okla.: University of Oklahoma, 1975).

64. "The Coal Industry's Controversial Move West," *Business Week*, May 11, 1974.

65. Ford, *op. cit.*

66. Rose, David J., *et al., op. cit.*

67. Smyth, Kirk R., John Weyant, John P. Holdren, *Evaluation of Conventional Power Systems,* ERG-75 (revised) (Berkeley, Calif.: Energy and Resources Group, University of California, Berkeley, 1975).

68. *Ibid.*

69. *Ibid.*

Part III

THE RISING TIDE
CITIZEN ACTION

13

THE RISING TIDE

CITIZEN ACTION

"If you construct something foolproof, there will always be a fool greater than the proof."
— Dr. Edward Teller, speaking at the Nuclear Energy Forum, San Luis Obispo, Calif., October 17, 1975.

"Nuclear power is a sword of Damocles. Scientists thrill in creating such things that take all man's ingenuity to control."
— Dr. David Lenderts, a primary organizer of the forum, October 18, 1975.

A powerful national mass movement of well-organized and well-informed nuclear opponents has now emerged in the U.S. to challenge the atomic industrial complex on political, economic, legal, and environmental grounds. From a small core of opponents in the late 1960s, this opposition has swelled to a broadly based citizens' movement today, with strongholds of support in the middle class (especially among professionals and women), in the academic community and, to an ever growing extent, among working-class people. The movement has been energized by a growing public consciousness of the direct relation between nuclear radiation, environmental pollution, and many public health hazards. In addition to these factors, the public's new critical attitude toward nuclear fission stems from —

(1) widespread dissatisfaction with spiraling utility rates, prompting a new, across-the-board skepticism about utility decisions of all kinds,

(2) increasing resentment toward big corporations in general,

(3) growing dissatisfaction with U.S. energy policy, among politically aware people,

(4) increasing knowledge of safe energy alternatives, and

(5) growing recognition of the basic connection between nuclear power and the U.S. failure to develop a sound energy policy based on clean alternatives.

The national anti-nuclear movement is linked by a strong network formed of several well-established environmental, research, citizens, and activist groups with large, influential memberships and with numerous publications devoted to energy or nuclear power information. Two new groups which officially entered the lists on the opponents' side in 1975 were the Americans for Democratic Action and Common Cause, which announced in April that it would oppose nuclear power plant construction until the plants' safety is established. As mentioned in Chapter 7, the National Council of Churches in 1976 also joined the nuclear opposition by taking a stand against the use of plutonium in reactor fuel.

As nuclear opponents' mass support has increased, the movement has become self-confident and outspoken, while showing signs of a new willingness to confront the nuclear complex directly, even if this confrontation requires personal sacrifices, risks, and expenses.

An indication of this new mood came at the Critical Mass's 1975 conference of nuclear opponents in Washington, D.C., where a representative of the South Carolina citizens' movement declared that an effort would be made to block trucks bearing nuclear wastes from entering the nearly completed Barnwell, S.C., reprocessing facility. But to date, citizens' actions against nuclear power in the U.S., with a few exceptions, have largely been confined to challenges made through officially sanctioned channels: court actions, public education campaigns, complaints to government agencies, interventions at AEC license hearings, and efforts to control nuclear power through the legislative process. These efforts have secured the sympathy and backing of some influential people in media and government, and have alerted millions of people to the dangers of large-scale nuclear expansion. With the creation of a mass base of opposition to nuclear power the stage may now be set for a new phase of bolder challenges to the atomic establishment. Clues to the direction U.S. protest may take are evident from this account of the recent struggle against nuclear industrial and governmental forces in West Germany.

The South Nuclear Power Plant Co. (Kernkraftwerk Sud) and its subsidiary (Baden Werk und Energieversorgung Schwaben) are collaborating with their French and Swiss industrial allies to build the densest concentration of nuclear reactors that the world has ever known, in the Upper Rhine Valley of West Germany. French and German citizen opponents united against the project in February 1975 for an action that may presage more determined resistance to the nuclear industry in the U.S. Local farmers and working class people risked arrest and physical injury by seizing and occupying a major nuclear power plant site and by defending it against thousands of police. Their dramatic confrontation brought an immediate halt to the plant's construction.

South Nuclear Power Plant Co. of West Germany had planned to build 1,350 MW of nuclear capacity at Wyhl in southwestern Germany — far more electricity than currently needed in the region. Surplus power was to be profitably exported, and new heavy, energy-intensive industries were to be attracted to the area. Citizens in the Wyhl-Weisweil area, famous for its Riesling and Liebfraumilch wines, became concerned about the plant's possible environmental effects. Humidity released by the plant's cooling towers would cause some fog and frost. What would the fog and frost do to the areas' vineyards and to the locally grown grains and tobacco? How would the area's new industries affect the Rhine, which provides drinking water for twenty-five million Europeans? What would the plant's thermal pollution do to the river?

Spurred on by these fears, local organization against the plant was well under way by the summer of 1974, before ground was even broken. To counter citizen efforts, the company and the government began trying to muster popular support for the plant.

State Minister President Hans Filbinger of the Baden-Wuettemberg state government and State Minister of the Economy Eberle both sat on the company's board of directors. (Minister Filbinger has since resigned that company post.) With Minister Filbinger's help, a referendum on whether the Wyhl site should be sold to South Nuclear was decided in the company's favor on January 22, 1975. No one but residents of Wyhl were allowed to vote in the referendum, although the plant clearly was to have a regional environmental impact.

Minister Filbinger authorized construction of the plant the day after the referendum results were in, and representatives of four neighboring communities and individuals promptly filed suit in a challenge to the referendum's legality and to the proposed construction. Plant opponents were particularly incensed that an environmental review of the project was to be conducted by experts chosen by the company, and that their work was to be subject to final approval by Minister Filbinger. Despite the pending court cases, the Wyhl plant site was sold to the company on February 1; construction began on February 17; and a significant citizens' protest was held there February 18.

According to People's Translation Service of Berkeley, California,* this is what happened next at the Wyhl site:

> Three hundred local residents attended the demonstration, watching the clearing of the site in shocked silence. After fifteen minutes, intense discussions with the construction workers began. The citizens then took control of the machines and the work stopped. An hour later, the police arrived and demanded that the demonstrators leave the area. Aware of their small numbers, the demonstrators retreated at first, but then when police reiterated their order, they returned to discuss the dangers of the nuclear reactor with police."

The demonstrators then decided to occupy the site and set up a tent there. They were soon joined by supporters from nearby communities in France and by local supporters who brought food, wine, and blankets. The occupiers, mostly farmers, set up security

---

*Peoples Translation Service provides English Translations of important foreign news stories on a regular basis to media publications and interested individuals. Their address is 2054 University Avenue, Berkeley, California. Many of the international nuclear protests recounted in this chapter recieved extensive coverage by PTS.

watches to warn of arriving police, and local residents built barricades on the site to prevent arrests. The protesters succeeded in holding the site for the remainder of that day and night, but the following day, strongly reinforced police units drove the farmers off with water hoses and clubs. Later, police surrounded the site with barbed wire. A rally held there two days later brought five thousand people to protest against the plant and, on February 23, more than twenty thousand area residents, with supporters from nearby Switzerland and France, again held a mass protest meeting at the site. After the meeting, farmers began tearing up the barbed-wire fences around the plant and routed three thousand police and attack dogs, to recapture the site.

Women from neighboring villages donated ample provisions, and the site was turned into a martial encampment. Access roads were barricaded with tree trunks, bonfires were lit, and everyone entering the site was screened to keep out police informants. About fifteen hundred people camped overnight on the spot and, as a result of the demonstrators' intransigence, a halt to plant construction was ordered until the court suits filed the previous month were resolved.

During the following eight months, many of the occupiers made the site their home. A dining hall was set up to serve the tent dwellers, and a variety of seminars, discussions, and study groups were held to inform local citizens about salient issues. A people's high school was also established. In one weekend seminar, the participants discussed reactor technology, alternative energy, and various other ecology issues.

Late in October 1975, representatives from a council formed by the farmers agreed temporarily to end the occupation while negotiations with the state government were conducted. At last report, construction remained halted, and further rulings in the German courts may take as much as two years, according to informed observers.

The Wyhl protests were not isolated events. Mass demonstrations were held against a number of nuclear power plants during the past five years in Western Europe. Notable among the demonstrations have been those at Fessenheim in Alsace, France. Large internationally attended annual gatherings have been held there since 1971 as part of the ongoing attempt to stop construction of a nuclear reactor currently being built there. And at Schweinfurt, West Germany, on April 19, 1975, ten thousand people marched in protest against the construction of a nuclear power plant at nearby Grafenrheinfeld. Many of the participants in the protest were farmers who brought their tractors with them. Opponents of the plant charged it would destroy the agriculture in the region by heating the River Main with its thermal discharges. They collected thirty-six thousand signatures on a petition demanding the plant be abandoned but a preliminary court injunction obtained in their favor was later rescinded by the Bavarian State Appellate Court which actually ruled that "a supply of electricity for the area is more important than the protection of life, health, and property."

Other major demonstrations were held during March and April in Kaiseraugst, a small town on the Rhine near Basel, Switzerland. Fifteen thousand people gathered on April 6, 1975, to protest the construction of a nuclear power plant there, and to support an ongoing site occupation. When excavation began on March 27, the Kaiseraugst site was occupied by more than five hundred people, and excavation was stopped.

Also during April 1975, farmers in Billigheim, West Germany, drove their plows and other agricultural equipment onto the site of their country's first regional nuclear waste depot. They prevented a construction crew from doing a test drilling and a militant who spoke for the occupiers threatened, "If only the slightest action is taken, all hell will break loose."

U.S. DIRECT ACTION

Few American groups or individuals have so far chosen to confront major nuclear power projects directly, with the interesting exception of Samuel H. Lovejoy, a resident of a farming commune near Montague, Massachusetts, and a political science graduate of nearby Amherst College. In an act of deliberate civil disobedience that has received wide publicity and would probably have earned the disapproval of Lovejoy's Amherst classmate, David Eisenhower, Lovejoy used a pipe wrench and wrecking bar to destroy a five hundred-foot tower erected by Northeast Utilities Co. to study weather conditions prior to construction by the company of twin reactors with 2,300 Mw of combined capacity on the Montague Plains.

After Lovejoy toppled the tower by loosening three turnbuckles at its base, he surrendered to local police with a seven hundred fifty-word statement denouncing the nuclear power plant.

Lovejoy was tried in Franklin County Superior Court in Greenfield, Massachusetts, on charges of "willful and malicious destruction of personal property." The felony charge carries a five-year maximum prison term in Massachusetts. Acting as his own counsel for most of the trial, Lovejoy called as his witnesses nuclear power expert Dr. John W. Gofman and radical historian Howard Zinn, a specialist on civil disobedience. Both were allowed to testify for the record rather than in the jury's presence, but their depositions and Lovejoy's own six-hour defense on the stand apparently convinced Judge Kent Smith that Lovejoy's motives were not "malicious." The judge instructed the jury to find Lovejoy "not guilty" on the ground that the property destroyed was "real" and not "personal." [1]

Despite his acquittal, Lovejoy reportedly was disappointed that the case was not resolved on the issue of nuclear power safety. Yet his civil disobediance served as a rallying point for nuclear opponents in Massachusetts, and the action received nationwide publicity.

So far, little of the protest against nuclear power in the U.S. has been expressed in the streets by means of mass marches or demonstrations. Those which have occurred were generally held in a restrained manner, reflective of the middle-class roots of the U.S. nuclear opposition movement. One example is the orderly mass demonstration called by Project Survival of Palo Alto, California, that brought more than four thousand people to the state capital, commemorating the death of plutonium technician Karen Silkwood. As related in Chapter 5, Silkwood was killed in a mysterious car accident while on her way to talk with a *New York Times* reporter about alleged unsafe conditions and alleged falsification of safety reports at the Kerr-McGee plutonium fuel fabrication facility.

LEGISLATIVE ACTION AND INTERVENTIONS

More typical of American efforts to stop nuclear power have been various attempts to use legislative or administrative processes to curb the nuclear industry. Tactics have ranged from interventions at AEC power plant hearings to initiatives and referenda.* Some environmental and citizens' groups have also been successful in securing state and citywide legislation or resolutions on nuclear power; others have urged administrative agencies and executives to promulgate new regulations for the nuclear industry.

Because some state legislators, with powerful industrial and financial connections, are notably reluctant to oppose nuclear power, many nuclear opponents have opted to use initiatives to circumvent the legislature.† But citizen activists in Vermont and New York have demonstrated that if strong popular and organizational pressures are exerted on a legislature, those bodies can greatly aid the nuclear opponents' cause.

In New York State, a coalition of more than a hundred environmental organizations coordinated by citizen activists Pauline Davis and Shirly Brand (and others) persuaded the New York state legislature to approve a Safe Energy Act late in June 1975. Although by the time the bill was passed, a nuclear power ban which the coalition had written into their draft legislation had been dropped by the lawmakers, the act had the important effect of reconstituting the state's Atomic and Space Development Authority into an energy research and development administration, raising non-nuclear energy alternatives to the same priority as nuclear power. This achievement may well slow nuclear power's development in New York and may assist the development of other energy options. Davis and Brand spent three to four days each week lobbying for the bill during the legislative session, supported by the state's united environmental movement.

Nuclear opponents scored another success in Vermont. The Vermont Public Interest Research Group won the state legislature's approval for a bill requiring legislative approval for the construction of any new reactors in the state. When the act was signed into law on April 3, 1975, Vermont became the first state to interpose the legislature as another tier of regulation in the nuclear power licensing process. Both houses of the legislature now must approve

---

*A referendum is a vote held giving the electorate an opportunity to approve or disapprove a law passed or proposed by a legislature.

†An initiative enables the people to propose and enact a law directly, without securing legislative approval.

a plant before the state's Public Utilities Commission can issue it a construction permit. An intensive lobbying and petition drive by the Vermont Public Interest Research Group was primarily responsible for the passage of this law.

Citizen action throughout the country in early 1976 began focusing on the issue of radioactive waste transport. Laws are being considered in New York City and Nassau County that would ban the transportation or storage of dangerous substances, including radioactive elements, except in emergencies and for biomedical reasons. Similar legislation is currently being considered in twenty-two states, one hundred fourteen cities, and more than three hundred counties.[2] If approved in Nassau County and nearby Suffolk County (where it has not yet been introduced), it would force the Long Island Lighting Company to abandon its plans for two nuclear power plants there, because disposal of the spent fuel would become impossible.

Various efforts have been made by nuclear opponents to restrict nuclear power expansion by intervening at AEC and now Nuclear Regulatory Commission hearings on plant construction permits and operating licenses. Although some of these interventions have generated useful publicity and public support, their effect has largely been to delay plant construction rather than avert it. Sometimes the interventions have obliged plant operators to modify plant designs. Important environmental concessions have been wrung from the nuclear industry at these proceedings, which take place largely according to AEC-set rules and regulations, but the monetary costs to intervenors have been high. Interventions so far have never resulted in the denial of a construction or operating permit and have often frustrated environmentalists because the cases are heard by an NRC panel sitting as both judge and jury.

A complex and fascinating controversy in which nuclear opponents lost a battle with Northern States Power Company in court and at a major intervention but then won a partial victory later when the power company capitulated to political pressure is described in detail by science writer Richard S. Lewis in *The Nuclear Power Rebellion, Citizens vs. the Atomic Establishment.*[3] During the first phase of this battle, the Northern States Power Company was eventually sustained by the U.S. Supreme Court in its fight against Minnesota's effort to impose state radiation standards on the company's Monticello, Minnesota, nuclear power plant. The state's standards were more stringent than the AEC's federal standards. In a decision that has since greatly handicapped environmentalists, the U.S. District Court in St. Paul ruled on March 17, 1971 that the federal government has the exclusive right to regulate the operation and construction of nuclear power plants under the Atomic Energy Act of 1954. The ruling held that the act pre-empted the right of a state to regulate nuclear power, including the regulation of radioactive discharges from power plants.

Before this issue was decided in the courts of Minnesota, intervention against the plant at the operating-license stage by graduate students in physics from the University of Minnesota and eight environmental groups had also met with defeat on September 8, 1970 when the AEC had granted the power company a limited license to operate, although the intervenors had challenged the plant's design specifications and the AEC's quality-control and radiation-regulation procedures. But then the activists mounted a campaign to influence the state legislature. Bills were introduced banning the construction of nuclear power plants and Governor Wendell Anderson, in a special message to the legislature, recommended passage of a nuclear-power moratorium. The next day, the Minnesota Pollution Control agency received a letter from the president of Northern States Power stating that pollution control devices would be installed at the Monticello plant, reducing gaseous radioactive releases by "at least 95 percent," as quoted in Lewis's account.

In another legal battle, environmentalists and representatives of various fishermen's groups intervened in 1970 against the 700 Mw Palisades Nuclear Power plant proposed for construction on Lake Michigan by Consumers Power Company. After months of effort at public hearings on the plant's operating license during which intervenors charged that the plant's radiation and heat discharges would threaten Lake Michigan with pollution the company yielded to the intervenors to avoid further delays in its licensing process. The firm opened direct negotiations with the intervenors, thereby giving them considerable *de facto* authority in regulating the plant's radioactive releases. The intervenors, in turn, agreed to drop their case provided that the company install closed-cycle cooling towers at the plant and a new liquid waste system to reduce radioactive liquids released to near zero. In addition, the company agreed to adopt gaseous-emission controls in the future, if technically and economically practical. The company estimated the changes would cost $11 million, and insisted that, for the agreement to become effective, it had to have the right to pass the added costs on to consumers by adding the extra investments into its rate base. In a similar case, Wisconsin Electric Power Company and its subsidiary found it expedient to agree to the installation of new radioactive emission controls at its Point Beach Nuclear Power Plant at Two Creeks, Wisconsin, in order to induce intervenors to withdraw their objections to the plant.

Other legislative action has included local referenda on nuclear power in several communities throughout the country and, more importantly, it has included the drive to halt nuclear expansion by the initiative process, described in Chapter 8. In addition to putting nuclear power initiatives on the ballot in California for the June 1976 election and in Oregon for the November 1976 election, initiative supporters in early 1976 were circulating petitions in Arizona, Colorado, Maine, Missouri, Montana, and North Dakota. Nuclear opponents were also getting ready to file petitions in seven other states. The initiative process was revived as a potent political tool by Joyce and Ed Koupal, founders of the People's Lobby in Los Angeles. The Koupals have been active in promoting the initiative efforts throughout the country, wherever state laws grant citizens the right to use initiatives.* The nuclear opposition movement lost a powerful ally, a robust campaigner, and a dynamic leader when Ed Koupal succumbed to cancer in early 1976 after having dedicated the last years of his life to the struggle for responsive government and a clean environment.

COURT ACTION

Until recently, the courts have rarely permanently stopped the

---

*The Koupals demonstrated the power of the initiative technique by reviving it from relative obscurity and using it to qualify California's Political Reform Initiative of 1974 (Proposition 9), which then won voter approval.

construction of a nuclear power plant.* But in April 1975, the Portage, Indiana, chapter of the Izaak Walton League, joined by Business and Professional People for the Public Interest, won what appeared to be a major victory for all nuclear opponents. As a result of their suit against the Northern Indiana Public Service Company's proposed Bailey Nuclear Power Plant on the shores of Lake Michigan, a three-judge federal court of appeals panel in Chicago issued a permanent stay against construction of the plant. The court ruled that the plant, located just two miles from Portage, was too close to populated areas and therefore violated the Nuclear Regulatory Commission's own siting regulations.[4] This decision was reversed by the U.S. Supreme Court on November 11, 1975 which ruled that the appeals court had erred in rejecting the AEC's initial interpretation of its own plant siting regulations.

Two important pieces of environmental legislation of the 1970s, the National Environmental Protection Act of 1970 and the Federal Water Pollution Control Act of 1972, have led to landmark court decisions on nuclear power. NEPA first became an important issue in the annals of nuclear-power-plant siting during a controversy over the Calvert Cliffs Nuclear Generating Station that was constructed during the late sixties and early seventies on Chesapeake Bay near Lusby, Maryland. This Baltimore Gas and Electric Co. project consisted of two 800 MW reactors which were to go on line with a year's delay, starting in 1973. (The first actually came on line in 1974.)

From the onset of the plant's construction permit hearing in 1969, the plant was a source of controversy. The Chesapeake Environmental Protection Association formed by local concerned citizens, intervened against the construction permit. CEPA produced expert witnesses who stated that tritium contamination from the plant could be a health hazard to those eating seafood from the bay, and that thermal pollution from the plant could pose a threat to the bay's marine life. The AEC granted a construction permit anyway and, on December 3, 1970, under pressure from citizen activists, the AEC promulgated a set of rules for applying

---

*One exception was the case in which Pacific Gas and Electric Co. of California was prevented from building a nuclear reactor on or near an earthquake fault at Point Arena, near Bolinas, California. The Company had already done extensive excavation when finally ordered to abandon the project.

NEPA to reactor sites. The rules exempted the Calvert Cliffs plant from the act, along with other plants that had already received construction permits before the act went into effect. The intervenors (aided by the Sierra Club and the National Wildlife Federation) formed a new organization and challenged the AEC in court. The U.S. Court of Appeals for the District of Columbia sustained them and applied NEPA to plants like Calvert Cliffs, for which operating permits had not yet been granted. The court also required the AEC to hold hearings on nonradiological as well as on radiological environmental impacts of power plants.[5]

About a year after NEPA went into effect, the Natural Resources Defense Council, a national environmental group with about twenty thousand members and contributors, sought to implement NEPA in a different way by attempting to force the AEC to apply the act to its breeder reactor program. Prior to the filing of the council's suit in May 1971, NEPA had been applied to force the assessment of the environmental impacts of individual projects, rather than of entire atomic energy programs. The Council's position was sustained by the U.S. Court of Appeals in June 1973, requiring the AEC to prepare an environmental impact statement for the breeder program. The AEC report was then duly prepared and submitted, as is customary, to other governmental agencies and outside experts for review and comment. But to the AEC's consternation, the U.S. Environmental Protection Agency's reaction was outspokenly negative. In an official analysis of the AEC document, the EPA termed the report "inadequate," and said it was "particularly deficient in its treatment of reactor safety, in potential problems associated with plutonium toxicity and safety and the cost benefit analysis."

The EPA also sharply criticized the AEC for failing to consider properly the merits of alternatives to breeder reactors, such as solar energy. EPA also urged that the AEC obtain a delay from the federal court to revise the report. The effect of the Natural Resources Defense Council's legal effort has been to require the Energy Research and Development Administration to prepare environmental impact reports for other major energy programs, such as the nuclear export program and the nuclear waste-management program. The threat of pointed and embarrassing criticisms from independent scientists and from other governmental agencies now acts as a check on the programmatic environmental assessments prepared by ERDA.

In another important application of national environmental legislation, the Colorado Public Interest Research Group (CoPIRG) and Colorado Environmental Legal Services, Inc. used the 1972 Water Pollution Control Act to force the U.S. Environmental Protection Agency to regulate the radioactive discharges from nuclear power plants. Previously, the responsibility for regulation rested with the AEC and then with the Nuclear Regulatory Commission, both of which were likely to be more lax in their standard-setting and enforcement than the EPA. The environmentalists' suit filed in October 1973 asserted that the federal act obliges the EPA to control radioactive discharges into U.S. waterways. The Tenth U.S. Circuit Court of Appeals agreed, overturning an earlier district court verdict, and also ordered the EPA to set national water-pollution standards for all U.S. nuclear facilities.[6]

REGULATORY AGENCIES

In addition to opposing nuclear power through the courts and by legislative means, citizen activists have also appealed to state public utility commissions and to three important federal regulatory agencies: the Federal Power Commission, the Federal Communications Commission, and the Federal Trade Commission.

One recent complaint filed with the Federal Power Commission by the Media Access Project on June 30, 1975, could have far-reaching consequences. Representing fifteen groups and individuals, Media Access Project of Washington, D.C., petitioned the Federal Power Commission to rule that nuclear power advertising by utilities was "political" advertising and therefore could not properly be included in a company's rate base. Such a ruling would severely inhibit this advertising because political expenses cannot be passed on to ratepayers and must be borne by the company's stockholders, in the case of investor-owned utilities.

In California, Project Survival, the citizens' group with headquarters in Palo Alto, filed a complaint with the state's Public Utilities Commission about the alleged unfairness of the Pacific Gas & Electric Company's nuclear advertising. (See Chapter 8.)

Complaints about utility advertising can also be lodged with the Federal Communications Commission and the Federal Trade

Commission. The Federal Communications Commission is responsible for enforcing the Communications Act of 1934, which requires broadcasters to provide reasonable opportunities for the presentation of opposing viewpoints on controversial issues. Under this "fairness doctrine," citizens can request equal time to reply to pro-nuclear utility advertising. The Public Media Center of San Francisco has provided nuclear opponents with hard-hitting anti-nuclear advertising for this purpose.

Protests against nuclear advertising have also been filed with the Federal Trade Commission. The FTC is obliged by the Federal Trade Commission Act to issue "cease and desist orders" if instances of false, unfair, or misleading advertising are brought to its attention. So far, the FTC has done little to enforce fair nuclear advertising.

A new and potentially forceful tool for citizens to use against nuclear power — and to assure a new accountability by utilities towards consumers — would be the use of Residential Utility Citizen Action Groups. RUCAG, an idea advocated by the Nader movement, would basically operate as a public advocate, utility ombudsman, and more. It would represent consumer interests in virtually all cases when those interests need protection from utilities of all kinds, such as telephone, water, gas and electric companies. A RUCAG, funded by a voluntary check-off box on utility bills, could be established by state law or by action of a public utilities commission.

The RUCAG would intervene at utility rate proceedings before the PUCs to protect consumers against high charges and rate increases. It would also handle consumer complaints; monitor political activity by the utilities; protest to federal regulators when necessary about arbitrary utility actions; and would inform and mobilize citizens on important issues affecting their well being.

If established as described above, the RUCAG would be accountable to the people, rather than to political officials. Moreover, to guard the organization against cooptation, no one with conflicts of interests—such as utility employees or stockholders—would be allowed to become a RUCAG director. One great advantage to consumers would be the RUCAG's ability to hire lawyers, accountants, engineers, and economists to challenge utilities' legal, technical, and economic assessments. Utility appeals to PUCs for rate

increases and justifications of utility policies are based on such
assessments. By virtue of its permanence, its economic links directly
to its popular constituency, and its political independence, a
RUCAG may build one of the most valuable and lasting political
structures to emerge from the nuclear opposition movement. Further
information about how RUCAGs can be set up and what they are
apt to accomplish can be obtained from the Center for Responsive
Law's *Citizens' Guide to Nuclear Power.*

PUBLIC EDUCATION

In the long run, effective opposition to nuclear power depends on
nuclear opponents' ability to build a large and aroused public
constituency. Nuclear opponents are currently laying that groundwork with a vast variety of public education techniques. These
have included conferences, forums, symposiums, workshops, and
rallies; petitions, statements of concern, and background information on energy; newspaper ads, displays and films; organized
letter writing campaigns; speakers bureaus; and the solicitation of
endorsements of anti-nuclear energy platforms from prominent
officials and groups. Not all the opposition has been so somber.
The San Francisco Mime Troupe in 1975 put on "Power Plays," a
well-recieved, satirical,anti-nuclear theatrical production, written
by troupe members. And in Oregon, the Energy Conservation
Organization of Eugene put on "Dr. Atomic's World Famous
Medicine Show," a play dramatizing public concern about ecology
and nuclear hazards

The use of petition drives has also proven to be a valuable
focus both for publicity and for public education. For example
the Task Force Against Nuclear Pollution of Washington, D.C.,
has not only received useful media attention and helped hundreds
of thousands of individuals make a small but symbolic commitment
by endorsing an energy petition, it has also registered growing
voter disenchantment with nuclear energy. The Task Force has so
far gathered more than three hundred thousand signatures to its
Clean Energy Petition which states, "I, the undersigned, petition
my representatives in Government to sponsor and actively support
legislation to: (1) develop safe, cost-competitive solar electricity
and solar fuels within ten years or less, and (2) phase out the

operation of nuclear power plants as quickly as possible." (See page 369.) Once the Task Force receives signed petitions, it sorts them by congressional district and delivers them to U.S. representatives, showing them the mounting public opposition to nuclear power in their districts.

Signature gathering also proved effective when twenty-three hundred scientists signed a petition expressing grave concern over the dangers of nuclear power.[7] Circulated by the Union of Concerned Scientists, the petition was delivered to the White House and Congress on August 6, 1975, the 30th anniversary of the Hiroshima atomic bombing. The message reached a number of major newspapers and undoubtedly influenced millions of people to adopt a more critical stance toward nuclear power. The petition's impact was enhanced by the endorsements of many prestigious scientists, including nine Nobel laureates.

Naturally, the number of signatures is not the only determinant of a petition's effectiveness. In the fall of 1975, ninety-five physicians of San Luis Obispo, California, organized by Dr. David Lenderts, signed a petition that expressed concern about the possible health effects of radiation from nuclear power. This "Statement of Concern" received intense local publicity and helped prepare the public and media for an important forum on nuclear energy which the doctors and Another Mother for Peace of San Luis Obispo held at California Polytechnic State University on October 16 and 17, 1975. San Luis Obispo is just 12 miles northeast of Diablo Canyon, site of the largest nuclear generating station west of the Mississippi, where two nuclear reactors costing nearly a billion dollars are being constructed. The first, with 1,131 Mw capacity, is scheduled to go on-line in August 1976, the second (1,156 Mw) is to follow a year later. Eventually, an additional two reactors are scheduled to be built on the site.

The Diablo Canyon nuclear plant has already been the focus of a controversy between citizen activists, led by (among others) Mrs. Sandy Silver and her husband, Dr. Gordon Silver, a physics professor at California Polytechnic University. The Silvers and the local Another Mother for Peace Chapter challenged Pacific Gas & Electric Company, builders of the plant, before the Nuclear Regulatory Commission and prevented the company from storing radioactive fuel on the plant's site in 1975 before an operating permit for the plant had been granted. This was done by forcing the company to admit, through the use of formal interrogatories, that fresh reactor fuel, which was to be stored in the reactor's spent fuel pool, could "go critical" if the rods touched each other along their entire length in the presence of water. Nearby to the spent fuel pool in which an initial core load of 193 fuel rods was to be stored are huge water storage tanks. If the tanks were to rupture in an earthquake, they could flood the pools, according to Dr. Silver, and the rods then might "go critical."

San Luis Obispo is certain to be the center of an extensive citizens-utility conflict over the Diablo Canyon plant because of the discovery in January 1975 of the Hosgris earthquake fault that stretches along the ocean floor for at least eighty miles and passes within three miles of the plant.8 Although licensing proceedings for the plant had been conducted in the late 1960s, and two earthquake faults were known, at distances of fifty and twenty miles from the plant, none of the faults was previously considered a threat to the reactors. Contrary to these early assumptions, U.S. Geological Survey geologists have now concluded the Hosgri Fault is active. Experts have stated that the new fault could produce a quake up to five times greater than the devastating shock which hit the San Fernando Valley of California in 1971, and they are currently not sure how such an upheaval would affect the plant.*

The plant's ability to withstand an earthquake is certain to be a source of major controversy when the Atomic Safety and Licensing Board of the Nuclear Regulatory Commission holds hearings in San Luis Obispo on the plant's operating license. Those hearings currently are scheduled for the spring of 1976, but the newly evaluated fault and scientific controversies over its significance are likely to cause at least half a year's delay in the plant's licensing and could require redesign and reconstruction of the plant.

---

*There is also scientific controversy about the length of the fault line, and consequently about the severity of earthquakes that are likely to occur along it. At least one scientist believes the Hosgri Fault to be two hundred miles long, rather than eighty.

CONCLUSION

Citizen activists are pressing the nuclear industry on every front: in the hearing room, in the courts, in Congress, in state legislatures, and in city councils as well as in the appointive bureaucracies of government at every level. The activists are not funded by big industrial or financial institutions. For the most part, they are supported by small individual contributions and by small-scale fund-raising activities. (These activities have included coffee hours, dinners, potlucks, bake sales, yard sales, craft sales, raffles, tournaments, even the sales of anti-nuclear T-shirts, calendars, and buttons.) The nuclear critics have proven as resourceful in raising money as in opposing nuclear power itself, knowing that no one tactic should be exclusively relied on to quash nuclear power.

The opponents have learned not to base their opposition on intervention efforts alone because at best these have delayed plants and wrung concessions from their builders, without stopping the plants. Concerned citizens also have learned not to rely exclusively on legislators in state houses where good bills drafted by nuclear opponents can be whittled down or gutted before final passage. Nor can citizens count on the support of the U.S. courts, as not a single plant has been directly stopped by legal action. The most promising tactic that citizens currently are using is the initiative and, according to *Reuters* (April 5, 1976) eighteen states have already scheduled ballots on nuclear power.

Citizens will also continue to oppose utility policies before state public utilities commissions, and nuclear opponents will try to unify their cause with that of utility-rate opponents, thereby consolidating the environmental and consumer movement into one powerful mass movement. In so doing, the nuclear opponents may present their cause as a positive campaign *for* safe energy, not just a "movement against . . . "— which nuclear proponents would then seek to portray as a reactionary animosity towards progress.

While the majority in the nuclear movement pursues its fundamentally establishmentarian opposition to nuclear power, some groups will probably begin using more radical tactics of opposition by demonstrating against nuclear power at waste reprocessing facilities and at diverse stages of the nuclear fuel cycle. Other opponents have already called for economic pressure on utilities by urging the observance of specified days of minimal energy use by consumers.

The entire nuclear fuel cycle depends on the acquiescence, if not the complicity, of the union leadership in the transportation industry, specifically the trucking, construction, and railroad workers unions. Some opposition groups may begin cultivating contacts with labor unions in the years to come. One strategy may be to persuade union leaders of the obvious fact that nuclear power will not provide significant numbers of jobs, (especially relative to other technologies) and to discuss nuclear power's health hazards with rank-and-file union members.

To avoid the danger that, with the resolution of the nuclear issue, certain groups acting as utility watchdogs today will fade away as the nuclear program is defeated, the establishment of local Residential Utility Citizen Action Groups throughout the country will provide nuclear opponents with permanent agencies designed not only to curtail the nuclear industry's growth, but to endure long enough to preside over its demise and dismemberment.

The only way nuclear power can flourish is if we grant it free rein. It cannot succeed if opposed by steadfast, aroused people who are informed, organized, and dedicated to safe energy and conservation.

NOTES

1. Harvey Wasserman, "Nuke Developers on the Defensive," *WIN* magazine, December 5, 1974.

2. "Wide Support Developing for Plutonium Transit Ban," *Critical Mass*, vol. 1, no. 10, January 1976.

3. Richard S. Lewis, *The Nuclear-Power Rebellion, Citizens vs. the Atomic Industrial Establishment* (New York, N.Y.: Viking Press, 1972).

4. "Court Blocks Plan for Nuclear Plant Near Lake Michigan," *Wall Street Journal*, April 2, 1975.

5. Lewis, *Op. Cit.*

6. Skip Laitner and others, *Citizens' Guide to Nuclear Power* (Washington, D.C.: Center for the Study of Responsive Law, 1975). Center for the Study of Responsive Law, Box 19367, Washington, D.C. 20036.

7. David Burnham, "2,300 Scientists Petition U.S. to Cut Construction of Nuclear Power Plants," *New York Times*, August 7, 1975.

8. David Perlman, "Safety of Atomic Plant Challenged," *San Francisco Chronicle*, January 15, 1976.

14

CONCLUSION

Although nuclear advocates have tried to convince the public that the country is "running short of energy," and should convert to nuclear fission as quickly as possible, they have not been candid with us. We are indeed using our oil and natural gas supplies rapidly and, at current consumption growth rates, these fuels may be largely depleted in a matter of decades. But *we have no crisis in energy supply*, as the chapters on solar and other alternative energies have shown. We have enough energy available from the sun, the wind, the oceans, and the nation's large coal reserves to provide far more energy than we need — now or in the foreseeable future. We live in a giant and virtually inexhaustible energy flux that is more than adequate for our needs. Yet we *do* have a real crisis — a crisis in energy policy, and a closely related environmental crisis.

A mainstay of current U.S. energy policy is the aggressive promotion of fission power, especially the breeder reactor. President Ford's call in 1975 for the operation of two hundred fission reactors by 1985 reflects the administration's full-speed-ahead approach to nuclear power. Promotion of this sort has led to atomic power proliferation throughout the U.S., despite the environmental risks, high costs, and pathetic energy yields. During this nuclear heyday, the nuclear industry has busily enriched itself. But while the industry swelled on plump government contracts and subsidies, the utility bills for U.S. consumers kept rising. Simultaneously, the government virtually ignored the steps that would have brought consumers some relief: ERDA has given clean alternative energies and conservation only token support. The public's need for power from safe energy sources at fair prices has gone unfulfilled.

Solar power and conservation have not been neglected through accident or oversight or simple ignorance. The U.S. energy bureaucracy is not responsive to citizens' needs because it is overly responsive to the pervasive pressures of the giant energy corporations that wield such dangerous influence on our government. Our current energy policy is an expression of the large energy corporations' desire not to upset the current economic-political status quo, which rests on rapid energy growth rates and extravagant energy use. Implicit in the threat that energy consumption might be curtailed is the threat that the more affluent sectors of society which now use a disproportionately high share of energy will be required to limit their consumption. Also implicit is the idea that large corporate profits will fall as many individuals and small firms build safe, decentralized energy systems. For this reason, energy policy reform is an important step toward a more pervasive political reform that would diminish the influence of big corporations upon our lives.

Before a U.S. energy policy that emphasizes conservation and clean technologies can be implemented, the nation must renounce the ill-advised expansion of nuclear fission. When the money-hungry uranium fission option is foreclosed, then adequate resources can be devoted to the development of safe alternatives.

A primary reason for nuclear's hasty spread has been the general lack of public knowledge about it. Although an erosion of public support for nuclear power is apparent, what we are *really* witnessing is *an erosion of ignorance* about nuclear fission. This energy technology, unlike others, evolved clad in tight security regulations; the public was shielded from the unpleasant knowledge of radiation hazards. So nuclear power never enjoyed the support of an *informed* public. But as ignorance has been replaced with knowledge about fission's dangers and alternatives to them, the public has begun turning, in justifiable anger, against fission. If the unholy alliance formed by ERDA and the large energy corporations also can be thoroughly exposed, then perhaps U.S. energy policy can be made to serve the majority instead of a privileged few.*

We cannot bring about the necessary massive changes in U.S. energy policy just by individually attempting to implement conservation on a personal scale or merely by trying to use safe energies in our daily lives while giant economic-governmental interests stall the development of safe, cheap energy. People must unite politically in effective groups to counter that institutional recalcitrance. In addition, any organizations to which we belong, especially labor unions eventually must be impelled to support safe energy policies, and to stand with informed citizens against uranium fission. Then one day, this wanton nuclear insult to the biosphere will be revealed to all as the miscarriage of technology and the misuse of science that it is. But here and there across the landscape of America, radioactive husks of defunct, entombed fission plants will stand as monuments to the folly of allowing a scientific elite beholden to self-serving corporations to make our "technical" decisions for us.

---

*Readers who are interested in an alternative energy policy may find the policy blueprint contained in the Citizens Energy Platform interesting. This eighty-page proposal for a humane and ecologically sound energy policy offers alternatives in ten major energy policy areas including energy pricing, conservation, electric utilities, nuclear energy, and alternative energies. The platform was prepared by ten national environmental, consumer, and anti-poverty citizen groups and is available for two dollars from the National Consumers Congress, 1346 Connecticut Avenue, N.W., Washington, D.C. 20036.

ACKNOWLEDGEMENTS

Many people generously gave their time and energy to assist me with this book. I particularly would like to thank the individuals and organizations listed below for providing information, criticism, advice, or other help; however, none of those listed is responsible for the contents of this book or its accuracy:

Barbara Ames, the Atomic Industrial Forum, James H. Anderson, Jr., Larry Bogart, David R. Brower, Alden Bryant, Paul J. Channell, David Dinsmore Comey, L. Douglas DeNike, David K. Dunaway, the U.S. Energy Research and Development Administration, Daniel F. Ford, James Frey, Friends of the Earth, John W. Gofman, Leo Goodman, Lester Gorn, Lawrence Grossman, Nancy Guinn, James G. Hanchett, John P. Holliday, Jim Harding, Carl J. Hocevar, Peter Polland, Morgan G. Huntington, Charles Hyder, Homer Ibser, William R. Kaysing, John Klemenic, Charles Komanoff, Edwin A. Koupal, Jeanne Lance, Terry Lash, Mike Lopez, Steven Mayer, Jeff P. Marks, Jane Morita, Natural Resources Defense Council, U.S. Nuclear Regulatory Commission, Myra O'Shaugnessy, People's Lobby, Roger Rapoport, Ann Roosevelt, Lee Schipper, John Ward Smith, Otto J. M. Smith, Kirk Smith Russell Stetler, Vincent Taylor, Edward Westcott, Earl Williamson, George Yadigaroglu.

To those above who are scientists, I am grateful for expert guidance in many specialized fields of energy research — in one case, for performing the basic research that is so essential to our understanding of radiation's hazards, and in another case, for helping create the benign alternative energy technologies that today can be substituted for nuclear power.

To the lay people who have read portions of this manuscript and offered editorial suggestions, I owe profound thanks for advice on how best to communicate a complex mass of facts in a clear and readable fashion.

I appreciate the kindness of my adversaries in the national energy debate for their willingness to share technical information and for the cooperation they extended to me in this research effort.

APPENDIX

Radionuclide Content of Light Water Reactor Fuel decayed 150 days

Nuclide	Concentration curies/ton	Nuclide	Concentration curies/ton
hydrogen-3	692	cesium-134	213,000
krypton-85	11,200	cesium-136	20.8
strontium-89	96,000	cesium-137	106,000
strontium-90	76,600	barium-140	430
yttrium-90	76,600	lanthanum-140	495
yttrium-91	159,000	cerium-141	567,000
zirconium-95	276,000	cerium-144	770,000
niobium-95	518,000	praseodymium-143	694
technetium-99	14.2	neodymium-147	51
ruthenium-103	89,100	promethium-147	99,400
ruthenium-106	410,000	samarium-151	1,150
rhodium-103$_m$	89,100	europium-152	11.5
cadmium-115$_m$	44.3	europium-155	6,370
antimony-124	86.3	terbium-160	300
tin-125	20	neptunium-239	17.4
antimony-125	8,130	plutonium-238	281
tellurium-125$_m$	3,280	plutonium-239	330
tellurium-127$_m$	6,180	plutonium-240	478
tellurium-127	6,110	plutonium-241	115,000
tellurium-129$_m$	6,690	americium-241	200
tellurium-129	4,290	curium-242	15,000
iodine-129	0.038	curium-244	2,490
iodine-131	2.17		

Source: ORNL-4451

Further References on Nuclear Power

Barry Commoner (ed.), *Energy and Human Welfare, Vol. 1, Social Cost of Power Production* (New Yosrk: Macmillan Information, 1975)

Background Report to the National Council of Churches of Christ in the U.S.A. in support of *The Plutonium Economy:* A Statement of Concern, September, 1975. National Council of Churches, 475 Riverside Drive, New York City 10027.

Nuclear Blackmail, Emergency Response Plan for the State of California, Phase 1: The Threat, Office of Emergency Services, State of California.

Report of the Liquid Metal Fast Breeder Reactor Program Review Group, ERDA-1 (Washington, D.C.: U.S. ERDA) January 1975.

Report on Nuclear Energy, Anticipation, No. 21, October 1975 (Edumenical Hearing on Nuclear Energy: A Report to the Churches, Sigtuna, Sweden, June 24-29, 1975).

Amory B. Lovins, Social and Ethical Aspects of the Liquid-Metal Fast Breeder Reactor, Paper presented to the symposium, "Key Questions About the Fast Breeder," Technische Nogeschool, Delft (Netherlands), November 27-28, 1975.

Scientists Institute for Public Information, *Nuclear Power: Economics and the Environment* (6052 Claremont Avenue, Oakland, Calif. 94618, $2.)

Statistical Data on the Uranium Industry, U.S. ERDA, Grand Junction, Colorado, January 1, 1975.

Additional Reference on Fossil Fuel

National Academy of Science, National Academy of Engineering, National Research Council, *Air Quality and Stationary Source Emission Control,* A Report by the Commission on Natural Resources, Prepared for the Committee on Public Works, Serial No. 94-4, March 1975. USGPO, Washington, D.C. 20402, $8,60,

Further References on Citizen Action

Skip Laitner and others, *Citizens' Guide to Nuclear Power* (Washington, D.C.: Center for the Study of Responsive Law, 1975) Center for the Study of Responsive Law, Box 19367, Washington, D.C. 20036.

Terry R. Lash, John E. Bryson, and Richard Cotton, *Citizens' Guide: The National Debate on the Handling of Radioactive Wastes from Nuclear Power Plants,* (Palo Alto, Calif.: Natural Resources Defense Council, Inc.). Natural Resources Defense Council, Inc., 664 Hamilton Avenue, Palo Alto, Calif. 94301, $2.50

Dorothy Nelkin, *Nuclear Power and its Critics, The Cayuga Lake Controversy* (Ithaca, N.Y.: Cornell University Press, 1971). $1.75

Richard Morgan and Sandra Jerabek, *How to Challenge Your Local Electric Utility:* A Citizen's Guide to the Power Industry (Washington, D.C.: Environmental Action Foundation, 1974) Environmental Action Foundation, 720 Dupont Circle Bldg., Washington, D.C. 20036, $1.50.

The Directory of Nuclear Activists (Denver, Colo.: Environmental Action of Colorado, 2239 East Colfax Ave., Denver, Colo. 80206. $7 for non-profit environmental groups and nuclear-activist individuals and organizations; $25 for all others.

The California Nuclear Safeguard Initiave — The First Of It's Kind

THE PEOPLE OF THE STATE OF CALIFORNIA DO ENACT AS FOLLOWS:
Sec. 1. Title 7.8 (commencing with Section 67500) is added to the Government Code, to read:

TITLE 7.8. LAND USE AND NUCLEAR POWER LIABILITY AND SAFEGUARDS ACT

67500. This title shall be known and may be cited as the Nuclear Safeguards Act.

67501. The people and the State of California hereby find and declare that nuclear power plants can have a profound effect on the planning for, and the use of, large areas of the state, as do related facilities connected with the manufacture, transportation, and storage of nuclear fuel, and the transportation, reprocessing, storage, and disposal of radioactive materials from nuclear fission power plants.

67502. The people further find and declare that substantial questions have been raised concerning the effect of nuclear fission power plants on land use and land use planning, as well as on public health and safety. Such questions include, but are not limited to, (a) the reliability of the performance of such plants, with serious economic, security, health, and safety consequences; (b) the reliability of the emergency safety systems for such plants; (c) the security of such plants, and of systems of transportation, reprocessing, and disposal or storage of wastes from such plants from earthquakes, other acts of God, theft, sabotage, and the like; (d) the state of knowledge regarding ways to store safely or adequately dispose of the radioactive waste products from nuclear fission power plants and related facilities; and (e) the creation by one generation of potentially catastrophic hazards for future generations.

67503. A nuclear fission power plant and related facilities may be a permitted land use in the State of California and its waters and considered to be reasonably safe and susceptible to rational land use planning, and may be licensed by state or local agencies, and may be constructed in the state only if all of the following conditions are met:

(a) after one year from the date of the passage of this measure, the liability limits imposed by the federal government have been removed and full compensation assured, either by law or waiver, as determined by a California court of competent jurisdiction and subject to the normal rights of appeal, for the people and businesses of California in the event of personal injury, property damage, or economic losses resulting from escape or diversion of radioactivity or radioactive materials from a nuclear fission power plant, and from escape or diversion of radioactivity or radioactive materials in the preparation, transportation, reprocessing, and storage or disposal of such materials associated with such a plant; and

(b) after five years from the date of the passage of this measure

(1) the effectiveness of all safety systems, including but not limited to the emergency core cooling system, of any nuclear fission power plant operating or to be operated in the State of California is demonstrated, by comprehensively testing in actual operation substantially similar physical systems, to the satisfaction of the Legislature, subject to the procedures specified in Section 67507; and

(2) the radioactive wastes from such a plant can be stored or disposed of, with no reasonable chance, as determined by the Legislature, subject to the procedures specified in Section 67507, of intentional or unintentional escape of such wastes or radioactivity into the natural environment which will eventually adversely affect the land or the people of the State of California, whether due to imperfect storage technologies, earthquakes or other acts of God, theft, sabotage, acts of war, governmental or social instabilities, or whatever other sources the Legislature may deem to be reasonably possible.

67504. (a) If within one year from the date of the passage of this measure the provisions of subsection 67503(a) have not been met, then each existing nuclear fission power plant and such plants under construction failing to meet the conditions specified in subsection 67503(a) shall not be operated at any time at more than sixty per cent of the original licensed core power level of such plant.

(b) Beginning five years from the date of the passage of this measure, each existing nuclear fission power plant and each such plant under construction shall not be operated at any time at more than sixty per cent of the licensed core power level of such plant and shall thereafter be derated at a rate of ten per cent per year of the licensed core power level of such plant, and shall not be operated at any time in excess of such reduced core power level, unless all of the conditions enumerated in Section 67503 are met.

67505. The provisions of Sections 67503 and 67504 shall not apply to small-scale nuclear fission reactors used exclusively for medical or experimental purposes.

67506. One year from the date of the passage of this measure, the Legislature shall initiate the hearing process specified in Section 67507, and, within three years from the date of the passage of this measure, determine whether it is reasonable to expect that the conditions specified in Section 67503(b) will be met. Unless the Legislature determines that it is reasonable to expect that the conditions of Section 67503(b) will be met, then nuclear fission power plants shall be a permitted land use in California only if such existing plants and such plants under construction are operated at no more than sixty per cent of their licensed core power level. Unless the determinations specified in this section are made in the affirmative, then neither the siting nor the construction of nuclear fission power plants or related facilities shall be a permitted land use in California.

67507. The determinations of the Legislature made pursuant to subsection 67503(b) and Section 67506 shall be made only after sufficient findings and only by a two-thirds vote of each house.

(a) To advise it in these determinations, the Legislature shall appoint an advisory group of at least fifteen (15) persons, comprised of distinguished experts in the fields of nuclear engineering, nuclear weaponry, land use planning, cancer research, sabotage techniques, security systems, public health, geology, seismology, energy resources, liability insurance, transportation security, and environmental sciences; as well as concerned citizens. The membership of this advisory group shall represent the full range of opinion on the relevant questions. The group shall solicit opinions and information from responsible interested parties, and hold widely publicized public hearings, after adequate notice, in various parts of the State prior to preparing its final report. At such hearings an opportunity to testify shall be given to all persons and an opportunity to cross-examine witnesses shall be given to all interested parties, within reasonable limits of time. The advisory group shall make public a final report, including minority reports if necessary, containing its findings, conclusions, and recommendations. Such report shall be summarized in plain language and made available to the general public at no more than the cost of reproduction.

(b) To ensure full public participation in the determinations specified in subsection 67503(b) and Section 67506, the Legislature shall also hold open and public hearings, within a reasonable time after the publication of the report specified in subsection (a) of this section, and before making its findings, giving full and adequate notice, and an opportunity to testify to all persons and the right to cross-examine witnesses to all interested parties, within reasonable limits of time.

(c) All documents, records, studies, analyses, testimony, and the like submitted to the Legislature in conjunction with its determinations specified in subsection 67503(b) and Section 67506, or to the advisory group described in subsection (a) of this section, shall be made available to the general public at no more than the cost of reproduction.

(d) No more than one-third of the members of the advisory group specified in this section shall have, during the two years prior to their appointment to the group, received any substantial portion of their income directly or indirectly from any individual, association, corporation, or governmental agency engaged in the research, development, promotion, manufacture, construction, sale, utilization, or regulation of nuclear fission power plants or their components.

(e) The members of the advisory group shall serve without compensation, but shall be reimbursed for the actual and necessary expenses incurred in the performance of their duties to the extent that reimbursement is not otherwise provided by another public agency. Members who are not employees of other public agencies shall receive fifty dollars ($50) for each full day of attending meetings of the advisory group.

(f) The advisory group may:

(1) Accept grants, contributions, and appropriations;

(2) Create a staff as it deems necessary;

(3) Contract for any professional services if such work or services cannot satisfactorily be performed by its employees;

(4) Be sued and sue to obtain any remedy to restrain violations of this title. Upon request of the advisory group, the State Attorney General shall provide necessary legal representation.

(5) Take any action it deems reasonable and necessary to carry out the provisions of this title.

(g) The advisory group and all members of the advisory group shall comply with the provisions of Sections 87100 through 87312 inclusive, of Title 9 of the California Government Code.

(h) Any person who violates any provision of this section shall be subject to a fine of not more than ten thousand dollars ($10,000), and shall be prohibited from serving on the advisory group.

67508. (a) The Governor shall annually publish, publicize, and release to the news media and to the appropriate officials of affected communities the entire evacuation plans specified in the licensing of each nuclear fission power plant. Copies of such plans shall be made available to the public upon request, at no more than the cost of reproduction.

(b) The Governor shall propose procedures for annual review by state and local officials of established evacuation plans, with regard for, but not limited to such factors as changes in traffic patterns, population densities, and new construction of schools, hospitals, industrial facilities, and the like. Opportunity for full public participation in such reviews shall be provided.

Sec. 2. There is hereby appropriated from the General Fund in the State Treasury to the legislative advisory group created by Section 67507 of the Government Code the sum of eight hundred thousand dollars ($800,000) for expenditures necessary in carrying out the responsibilities and duties set forth in Section 67507 of the Government Code.

Sec. 3. Amendments to this measure shall be made only by a two-thirds affirmative vote of each house of the Legislature, and may be made only to achieve the objectives of this measure.

Sec. 4. If any provision of this measure or the application thereof to any person or circumstances is held invalid, such invalidity shall not affect other provisions or applications of the measure which can be given effect without the invalid provision or application, and to this end the provisions of this measure are severable.

A CLEAN ENERGY PETITION

1. I, the undersigned, petition my representatives in Government to sponsor and actively support legislation to: (1.) develop safe, cost-competitive solar electricity and solar fuels within ten years or less, and (2.) phase out the operation of nuclear power plants as quickly as possible.

Signature	Name printed clearly	Date

Street Address (students: where you vote)	City	State	Zip

2. I, the undersigned, petition my representatives in Government to sponsor and actively support legislation to: (1.) develop safe, cost-competitive solar electricity and solar fuels within ten years or less, and (2.) phase out the operation of nuclear power plants as quickly as possible.

Signature	Name printed clearly	Date

Street Address (students: where you vote)	City	State	Zip

3.

I, the undersigned, petition my representatives in Government to sponsor and actively support legislation to: (1.) develop safe, cost-competitive solar electricity and solar fuels within ten years or less, and (2.) phase out the operation of nuclear power plants as quickly as possible.

| Signature | Name printed clearly | Date |

| Street Address (students: where you vote) | City | State | Zip |

4.

I, the undersigned, petition my representatives in Government to sponsor and actively support legislation to: (1.) develop safe, cost-competitive solar electricity and solar fuels within ten years or less, and (2.) phase out the operation of nuclear power plants as quickly as possible.

| Signature | Name printed clearly | Date |

| Street Address (students: where you vote) | City | State | Zip |

PLEASE MAIL SIGNED PETITIONS TO:

Task Force Against Nuclear Pollution, INC.
153 E Street, S.E.
Washington, D.C. 20003

THE TASK FORCE IS HAPPY TO SUPPLY YOU with extra petitions. For the names and addresses of helpers in your area, send a stamped self-addressed envelope to the Task Force. When the Task Force receives signed petitions, *we sort them* by *Congressional District*, and we take them to the right Representatives. Not just once, but again and again. We *prove* that concern is growing in their own Districts. IT WORKS! In areas where there are already enough Clean Energy Petition-signers to tip an election, elected officials at *all* levels are really paying attention. THIS IS NO ORDINARY PETITION-DRIVE. The Task Force maintains a permanent registry of petition-signers, sorted by District, legible, and computerized. Signatures are never lost; they work for you repeatedly. **This nationwide petition-drive, which has been endorsed by Ralph Nader, is making nuclear and solar energy into major political issues. Donations needed .**

The Nuclear Energy Reappraisal Act

H.R. 4971

Be it enacted by the Senate and House of Representatives of the United States of America in Congress assembled, That this Act may be cited as the "Nuclear Energy Reappraisal Act".

STATEMENT OF FINDINGS AND PURPOSES

SEC. 2. (a) The Congress finds that there is serious division in the general citizenry and in the scientific community about the wisdom of a commitment to a further expansion of nuclear fission power because of: unresolved questions about the safety of nuclear plants; the potential danger to society from the use of special nuclear materials, such as plutonium, which if diverted from their intended uses, may be used as weapons of terror; the unresolved problem of the storage of nuclear waste materials for 250,000 years; and because the economic feasibility and reliability of nuclear plants continues to be in question.

(b) The Congress therefore declares that—

(1) the further deployment of civilian nuclear fission plants is inconsistent with the national security and public safety as required by section 3(d) of the Atomic Energy Act of 1954;

(2) the serious safety and environmental problems associated with nuclear fission power should be resolved before a further commitment to nuclear power is made by the United States Government; and

(3) the Office of Technology Assessment should undertake a comprehensive review of the safety, environmental, and economic consequences of the proliferation of nuclear fission plants in the United States and abroad in a completely independent manner.

CESSATION OF LICENSING OF NUCLEAR PLANTS

SEC. 3. (a) The Nuclear Regulatory Commission is directed to cease, beginning on the first day after the date of the enactment of this Act, the granting of licenses or limited work authorization for the construction of nuclear fission powerplants and the granting of licenses for the export of nuclear fission powerplants.

(b) This termination shall continue until the Congress, after having adequate time to study the results of the investigation described in section 4, shall make a determination that—

(1) the effectiveness of all safety systems, including but not limited to the emergency core cooling system, of any nuclear fission powerplant operating or to be operated in the United States is demonstrated by comprehensive testing, in actual operation, substantially similar physical systems, to the satisfaction of the Congress;

(2) the radioactive wastes from such a plant can be stored or disposed of, with no reasonable chance, of intentional or unintentional escape of such wastes or radioactivity into the natural environment to immediately or eventually adversely affect the land or the people of the United States, whether due to imperfect storage technologies, earthquakes or other acts of God, theft, sabotage, acts of war, governmental or social instabilities, or whatever other sources the Congress may deem to be reasonably possible;

(3) the effectiveness of security systems throughout the fuel cycle is demonstrated to the satisfaction of the Congress; and

(4) after analysis of all the safety, environmental, and economic consequences enumerated in section 5, nuclear fission plants are clearly superior to other energy sources, including renewable energy sources.

(c) This termination shall continue until the Congress, after having adequate time to study the results of the investigation described in section 4, shall provide by law—

(1) for resumption of the licensing of nuclear fission powerplants and the development of criteria and standards for the licensing of such plants; or

(2) that resumption of such licensing be permitted but only under limited conditions specified in that law.

(d) Beginning five years and 180 days after the date of the enactment of this Act, if the Congress has not determined, under section 3(b), that the licensing of fission plants may continue, each existing nuclear fission powerplant and each such plant under construction shall not be operated at any time at more than sixty percent of the licensed core power level of such plant and shall thereafter be derated at a rate of ten percent per year of the licensed core power level of such plant, and shall not be operated at any time in excess of such reduced core power level.

(e) The provisions of section 3 shall not apply to small-scale nuclear fission reactors used exclusively for medical or experimental purposes.

OFFICE OF TECHNOLOGY
ASSESSMENT STUDY

SEC. 4. (a) The Office of Technology Assessment is directed to undertake a comprehensive study and investigation of the entire fuel cycle from mining through fuel reprocessing and waste management and, as described in section 5, to determine the safety and environmental hazards of this cycle.

(b) The Office of Technology Assessment shall conduct this study independently. The Office in conducting the study shall request, receive and consider the comments and opinions of independent scientists, engineers, consumer, and environmental representatives. The Office shall hold informal public hearings on each major area of inquiry to permit interested persons to present information orally, to conduct or have conducted cross examination of such persons as the Office determines appropriate for a full airing of the issues and to present rebuttal arguments. A verbatim transcript shall be taken of any oral presentation in cross examination and shall be published by the Office. The Office shall have the power to enter into contracts with individuals or corporations for the purposes of conducting the study, but shall not enter into contracts with or rely primarily on the expertise of any industry or company which provides materials, management capabilities, research, or consultant services for nuclear fission power-plants or which otherwise in the judgment of the Office might have an interest in perpetuating the nuclear industry.

(c) All Government agencies shall cooperate to the fullest extent with the Office and shall provide access to their personnel and data. At the request of the Office, any Government agency shall furnish any information which the Office deems appropriate for the purpose of conducting the study. The Office is further empowered to compel the delivery of any information in the possession of the Nuclear Regulatory Commission, National Laboratories, or any person, corporation, or association which the Office deems necessary for conducting the study.

OFFICE OF TECHNOLOGY
ASSESSMENT REPORTS

SEC. 5. (a) Five years after the date of the enactment of this Act, the Office of Technology Assessment shall submit a final report to the Congress and the public concerning the safety and environmental hazard of nuclear fission powerplants and the nuclear fuel cycle.

(b) The Office will provide an annual report to the Congress and the public on the progress of the study, and provide the opportunity for an annual public hearing concerning the progress of the study. In each annual report the Office shall inform the Congress of the actions it has taken to fulfill the requirements of the Act, whether it has found any evidence that any persons have violated the laws and regulations relating to safety in the development or use of nuclear power or special nuclear materials, whether it has any evidence that the agencies of the Federal Government, present or past, which have the responsibility for insuring the safety of the nuclear fission power have not faithfully or effectively exercised their responsibilities, the extent to which other Federal agencies have cooperated with the Office, whether all information requests of the Office under section 4(c) have been complied with, whether and to what extent the Office has made provision to insure that all viewpoints have been adequately considered. The Office, in its annual report, shall also make available to Congress and the public, any information relating to the safety of the nuclear fuel cycle which has heretofore not been public information either because it was not publicly available, it was not completed in an analytical form, or for other reasons.

(c) The final report shall include recommendations as to whether a resumption of the licensing of nuclear fission power plants should be allowed, and if so, the conditions under which licenses should be granted. The report shall consider the following issues:

(1) The safety and environmental hazards associated with the entire nuclear fuel cycle, including, but not limited to the significance of frequent malfunctions in components of emergency core cooling systems as evidenced by the 166 abnormal occurrences reported by the Atomic Energy Commission in 1973, the six failures of emergency core cooling systems in semi-scale tests, and the significance of the British government's rejection of the light water reactor because of the danger of pressure vessel rupture.

(2) The short-term and long-term genetic effects of low level radiation.

(3) The economic implications of a commitment to nuclear fission powerplants, particularly in relation to:

(A) the long-term cost and availability of raw materials in light of the existence of a foreign uranium cartel;

(B) the cost implications of the frequent shut-down of nuclear plants including the costs of shut-down and start-up, inspections, the cost to consumers of purchase of alternative power during shut-down, unemployment benefits and other costs of unemployment that result from shut-downs;

(C) the economic wisdom of a commitment to an energy technology in which prudent safety management requires that all plants of similar design be shut-down when a serious safety problem arises at one plant, or sabotage of one plant is threatened; and the costs of necessary safeguards, including the costs of the design of the components of a nuclear transportation system, the costs, both public and private, for personnel, equipment and property to protect the projected 1,000 nuclear plants, reprocessing facilities and the thousands of components of the nuclear transportation system and the costs of decommissioning existing nuclear fission power-plants; and

(D) the total savings to nuclear plant operators arising from the subsidiaries to nuclear power by the Federal Government since the inception of the civilian nuclear power program including research costs, for programs such as the liquid metal fast breeder reactor, waste storage costs, regulatory costs, promotional costs, enrichment costs, safeguard costs, insurance subsidies through the Act commonly called Price-Anderson Act and any other costs associated with the development of civilian nuclear power.

(4) The storage of high level radioactive wastes which may remain dangerous for 250,000 years.

(5) The central question of proliferation, nationally or internationally, of nuclear fission powerplants in relation to possibly safer and cheaper alternatives, especially renewable energy sources.

(6) An assessment of whether utilities, as institutions, are financially and technically capable of operating nuclear plants safely in light of the high costs of safety measures and the 861 AEC documented abnormal events in utility operated nuclear plants in 1973.

(7) An assessment of the licensing processes of the Atomic Energy Commission (and its successor the Nuclear Regulatory Commission) which have permitted nuclear plants to be built over geologic faults and in other unsafe locations and which have allowed the continuation of license for utilities which have shown gross negligence in the operation and construction of nuclear plants.

PUBLIC INFORMATION

SEC. 6. (a) The Office of Technology Assessment shall be subject to section 552 of title 5, United States Code. The provisions of paragraphs (4) and (5) of section 552 of title 5, United States Code, shall not be construed to apply to any records of the Office which relate to the development, operation, or efficacy of the safety systems throughout the entire nuclear fuel cycle, except that provisions of paragraph (4) of such section may be construed to apply to such records in any case where the Office, after notice and opportunity for an agency hearing on the record, determines that such disclosure would result in irreparable injury to the competitive position of the person from whom the information was obtained. Such determination shall be subject to judicial review pursuant to section 552 of title 5, United States Code.

(b) As used in subsection (a)—

(1) the term "records of the Office" includes any application, document, study, report, correspondence, or other material or information or any part thereof received by or originated by the Office in connection with its duties under this Act; and

(2) the term "Office" means the Office of Technology Assessment, or any entity therein.

COMPENSATION FOR PUBLIC

SEC. 7. The Office of Technology Assessment shall, pursuant to rules promulgated by it, provide compensation for travel costs, per diem expenses, and experts' fees, and other costs in consulting with the Office, pursuant to the Office's responsibility under section 3(b), to any person who—

(1) has or represents an interest (A) which would not otherwise be adequately represented in such consultation, and (B) whose views are necessary for a full assessment of nuclear power and alternatives pursuant to this Act; and

(2) who is unable to participate effectively in such assessment because such person cannot afford to pay the cost of travel, per diem expenses and expert witnesses.

AUTHORIZATION

SEC. 8. There is authorized to be appropriated for the study under section 4 the sum of $15,000,000 for each of the first five fiscal years beginning after the date of the enactment of this Act.

INDEX

abnormal events, *see* accidents
accidents: consequences, 63-66, 69; basis for predictions, 58, 69; core melt, 55; loss of coolant (LOCA), 57, 68; maximum credible, 53, 64, 69; transportation, 198;
— in nuclear power plants, 242; Browns Ferry, 50-52; Fermi Breeder, 55, 64; Soviet breeder, 94
actinides, 45
activation products, 45
Advisory Committee on Reactor Safeguards, 194
Aerojet Nuclear Company, 56, 70
Agreements for Cooperation, 222; *see also* Bilateral Agreements
Agricultural residues, 237, 243, 324
Allied-General Nuclear Services, 37, 38, 102, 142
alpha emissions, 47
alternative energies, 253-336
Americans for Democratic Action, 340
Americans for Energy Independence, 175
American Institute of Architects, 244, 245
American Nuclear Energy Council, 175
American Nuclear Society, 170, 171, 177, 216
American Public Power Association, 170, 172
American Telephone and Telegraph Co (ATT), 169
Anaconda Copper Co., 82
anaerobic digesters, 325
Anerson, James H., Jr. and Sr., 308
Another Mother for Peace, 355
Atlantic Richfield Hanford Co., 106
"Atoms for Peace", 219
Atomic Energy Ace (1954), 220, 348
Atomic Energy Commission (AEC), 170, 216; Division of Reactor Development and Technology, 170, 216
— Films, 190
— Headquarters, 70, 99, 100, 102

Argentina, 221
Argonne National Laboratory, 276
Atomic and Space Development Authority, 346
"atomic industrial complex", 165, 166, 194, 339
Atomic Industrial Forum, 123, 165, 166, 175, 178, 179, 181
Atomic Safety and Licensing Board, 356
Aspin, Senator Les, 193
Ayers, Russell W., 207

Babcock and Wilcox Co., 169, 220
Baden Werk und Energieversorgung Schwaben, 341
Bailey Nuclear Power Plant, 350
Baltimore Gas and Electric Co., 350
Bankers Trust Co. (N.Y.), 168
banks, 165-168
Barber, Richard J., Associates, 223
Barnes, Dr. Fredrick Q., 132, 133
Barnwell Reprocessing Facility, 87, 88, 102, 142
Batelle Memorial Institute, 112; Pacific Northwest Laboratory, 113
Bechtel Corporation, 141, 142, 169, 178
BEIR Report, 79, 91
Beck, R. W. and Associates, 148
Berg, Charles A., 241, 244
beta particle, radiation, 47
Bethe, Dr. Hans A., 129, 175, 315, 316
Billingheim, W. Germany, 344
bi-lateral agreements, 220
biological concentration process, 55, 109
birth defects, stillbirths, 157
black lung disease, 38, 326
Boeing Aerospace Corp., 70, 280
Boer, Dr. K. W., 272
Bohn, Hinrich, 325
boron, 43
Brand, Shirly, 346
Brazil, 221
breeder economy, 93, 139
breeder reactor, 70, 86, 91, 139, 149-151; accidents, 93, 94; ad, 184; construction costs, 139, 140; doubling time, 139; Phenix, 139; safety, 93, 139; research budget, 158; target for terrorists, 200; safety requirements, speed of, 150